Naturalists' Handbook

Ants

GARY J. SKINNER
GEOFFREY W. ALLEN

*With plates by Geoffrey W. Allen
and line drawings by Sophie Allington*

Pelagic Publishing
www.pelagicpublishing.com

Published by Pelagic Publishing
www.pelagicpublishing.com
PO Box 725, Exeter, EX1 9QU, UK

Ants
Naturalists' Handbooks 24

Series editors
S. A. Corbet and R. H. L. Disney

ISBN 978-1-907807-60-2 Paperback

Digital reprint edition of:
ISBN 0-85546-305-8 (1996) Paperback

Text © Pelagic Publishing 2015
Plates and key drawings © Geoffrey W. Allen 1996
Line illustrations © Sophie Allington 1996

All rights reserved. Apart from short excerpts for use in research or for reviews, no part of this document may be printed or reproduced, stored in a retrieval system, or transmitted in any form or by any means, electronic, mechanical, photocopying, recording, now known or hereafter invented or otherwise without prior permission from the publisher.

British Library Cataloguing in Publication Data
A catalogue record for this book is available from the British Library.

Contents

Editors' preface

Acknowledgements

1	Introduction	1
2	Biology of ants	7
3	Identification	38
	Key I Worker ants	40
	Key II Queen ants	47
	Key III Male ants	53
	Quick-check field key to common ants	58
	Notes on the commoner British species	60
4	Techniques	64
	Some useful addresses	74
	References and further reading	75
	Index	80

Editors' preface

Ants have long intrigued naturalists. Ants themselves, and their intricate interactions with other organisms such as plants and aphids, are tempting subjects for investigation. It is possible to read about their fascinating biology in some excellent books, such as *Ants* by Hölldobler and Wilson. But the reader inspired to investigate the ants in the garden may need more practical guidance. This book is designed to meet that need. We are fortunate to have two authors with complementary skills. Gary Skinner (chapters 1, 2 and 4) introduces the natural history of ants, their relationships with other organisms, and some methods of study, and highlights topics on which more research is needed. Geoff Allen has written and illustrated keys (chapter 3) and prepared plates (pls. 1–10) that will enable readers to name their ants with confidence. Gary Skinner's Quick-Check Key in chapter 3 gives a quick reminder of the commoner genera and species of ants, for beginners. We are glad to have drawings by Sophie Allington illustrating chapters 1, 2 and 4.

We hope this book will enable ants to receive more attention in the fieldwork of amateur naturalists as well as students at school and university, and will lead more people into the rich literature and the endless possibilities for research on ants.

SAC
RHLD
April 1995

Acknowledgements

GWA acknowledges the support of his wife Alison Mary Allen, and A.K. Wallace, E.R. Gunby and A.P. Hazledene during the preparation of the illustrations. He is grateful for constructive comments on the keys from B. Bolton, S. Hoy and the late J.C. Felton, and he thanks W.A. Foster for the loan of material from the collection of the Cambridge University Museum of Zoology.

GJS wishes to acknowledge the assistance of Stephen Hewitt at the Tullie House Museum in Carlisle, P.W.H. Flint and John Whittaker of the University of Lancaster, and Barry Bolton of the Natural History Museum. A very special mention must go to Sally Corbet for her patient and excellent editing during this long project. Lastly he thanks his wife, Ann Marie, without whose support this book would never have been completed.

We also express our gratitude to those publishers and artists who have kindly given us permission to use figures from other works. Figs. 1 and 2 are based on drawings by Sean Burgess, and reproduced by permission of Shire Publications Ltd. Fig. 3 is reproduced by permission of the Biological Records Centre. Figs. 7, 8, 13, 15 and 27 appear by permission of Blackwell Science Ltd. Fig. 4 is reproduced by permission of Professor Wilson. Fig. 11 is adapted from a drawing in M.V. Brian's *Social Insects*, and permission to use it was granted by Chapman and Hall. Fig 12 has been redrawn from a figure by Bert Hölldobler, with his permission. Figure 17 is redrawn from a figure by Hangartner and used by permission of Professor Autrum. Fig. 18 was drawn by Henry Disney and published in his Handbook for the Identification of British Insects (Disney, 1983). Figs. 25 and 26 are based on those published in *The Study of Ants* by S.H. Skaife and are reproduced by permission of Longman Group Ltd. Figs. 21 and 22 are from *Ladybirds* by M. Majerus and P. Kearns in the Naturalists' Handbooks series. Other figures have been drawn for the present publication either by Sophie Allington (to whom great thanks) or by the authors.

GWA
GJS

1 Introduction

Most of us know what an ant looks like. We can confidently recognise a particular insect as an ant, even after a casual glance. How? What are the features that make an insect an ant? In early school biology lessons we learn that adult insects have wings. Ants are an exception to this almost universal rule. In our rapid identification of an ant we notice the *lack* of wings. All ants are social and have a variety of different forms in their societies. Although only the workers are wingless, these are far more numerous and active than the winged reproductives and thus more likely to be seen by the casual observer.

Several other insects share this winglessness, but ants have other features which help to set them apart. All ants belong to one family (the Formicidae) within the massive order of insects, the Hymenoptera. This order is divided into two major groups, the Apocrita and the Symphyta. The fundamental distinguishing feature is that the Apocrita have a waist, whereas the Symphyta do not. Ants are Apocrita and thus are waisted. To distinguish ants from the few other insects that combine these two striking features, winglessness and a waist, we look for the other two ant characteristics: one or two scale-like or bulbous swellings or nodes on the waist; and elbowed antennae. Any insect with no wings, a waist with nodes and elbowed antennae is an ant (fig. 1).

Do we really see all this in an animal less than one centimetre long when we recognise an ant at a picnic spot? No – the feature that gives the game away is the sociality, reflected in the sheer numbers. The fascination of ants lies in their sociality, and it is the organisation of ant societies, so similar to our own and yet so different, that will be the main focus of this book.

Myrmecologists, people who study ants, agree that all ants belong to just the one family mentioned above, the Formicidae. This is further divided into a number of subfamilies. Table 1 summarises the classification and shows which subfamilies are represented in Britain. We have a very sparse ant fauna, with only four subfamilies out of a total world count of 11. When we look at species numbers, the situation is even more depressing for the aspiring myrmecologist. Out of an estimated world list of over 10,000 species, Britain boasts fewer than 50. Ants are fundamentally sun-lovers and our cold climate does not suit them. Nevertheless, a number of species survive well in our temperate latitudes and provide fascinating material for study. Some show behaviour almost as dramatic as that of the well known tropical army and driver ants, so beloved of horror story writers.

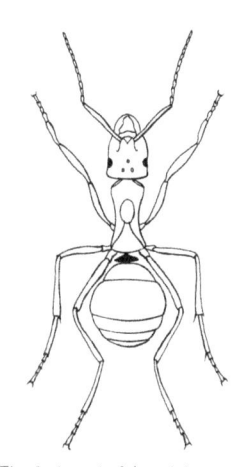

Fig. 1. A typical formicine ant, showing the one-segmented waist (after Skinner, 1987).

Table 1. *A world list of the subfamilies of ants (family Formicidae) (Hölldobler & Wilson, 1990). Asterisks mark subfamilies found in Britain. Bolton (1995) recognises more subfamilies than this.*

Sphecomyrminae (extinct)
*Ponerinae
Nothomyrmeciinae
Myrmeciinae
Dorylinae
Ecitoninae
Leptanillinae
Pseuodomyrmecinae
*Myrmicinae
Aneuretinae
*Dolichoderinae
Formiciinae (extinct)
*Formicinae

Two of the four British subfamilies are particularly poor in species. The Ponerinae are represented in Britain by only two, *Ponera coarctata* and *Hypoponera punctatissima*, neither of which is very common. *Ponera coarctata* is truly native and is found sporadically in the south. The status of *Hypoponera punctatissima* is the subject of debate, but it does now seem to be widely accepted that this species is not confined to greenhouses and hothouses. The features of the subfamily, and of the two British species, are given in chapter 3. The second small British subfamily, the Dolichoderinae, is represented by only four species, *Tapinoma erraticum*, *T. ambiguum*, *T. melanocephalum* and *Linepithema humile*. The first two of these are quite common and the last two are definite imports (known as 'tramp species').

Most of the ants we find are members of the two large British subfamilies, the Myrmicinae and the Formicinae. These are easily distinguished from one another with a hand lens or even sometimes without, once you have 'got your eye in'. We have seen that the waist of an ant bears scales or nodes. Formicine ants have just one (fig. 1) whereas myrmicines have two (fig. 2). It is useful to be able to distinguish these major groups in the field. Chapter 3 deals with identification in more detail.

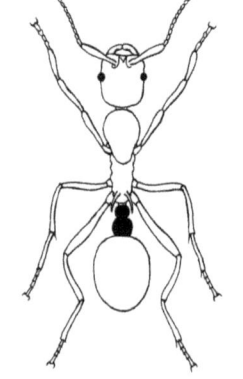

Fig. 2. A myrmicine ant, showing the two-segmented waist (after Skinner, 1987).

Where to look for ants

We have already seen that ants like warmth, and thus tend to be most abundant in warmer parts of the country (the south and west), or in warmer habitats (such as those on sandy soils).

In warm weather during the active season ants are easy to find, particularly when they are foraging for food. Thus, in woodland, it is always worth looking at the trunks of trees, on which workers may often be seen streaming up and down on their way to and from food sources amongst the leaves. In habitats dominated by shrubs and herbaceous plants, a careful search will often reveal ants looking for food.

Nearly all ants build nests and these may be more obvious than the ants themselves. The hill ant, *Lasius flavus*, is a largely subterranean species, and its 30 cm high, vegetation-covered mounds are much more likely to be spotted than the tiny, secretive workers. Nests of the wood ants (*Formica* species) are very large domes made of pine needles and other materials, often more than one metre high, and difficult to miss in the woodlands in which they occur.

Some species are rarely found. In some cases this is probably because they are truly rare. In others it is almost certainly due to want of looking; here is an area in which useful contributions could be made by amateurs. Very little ant recording has been done in many parts of the country, especially in the north (Barrett, 1979)* (fig. 3).

Some species are parasitic on others and do not form populous colonies. The distribution of such species (*Myrmica karavajevi*, *M. hirsuta*, *Anergates* and *Strongylognathus*) is almost certainly imperfectly known. *Myrmica karavajevi* occurs only in nests of *Myrmica scabrinodis* and *M. sabuleti*, and *Anergates* and *Strongylognathus* occur in *Tetramorium caespitum* nests. Careful searching for these could be well worth while. There is even a possibility that species hitherto unknown in Britain remain to be found. There is a suggestion that it would be worth looking, for example, for members of the genus *Doronomyrmex*, particularly in the Highland regions.

Other ants, although not parasitic in the same sense, are commonly associated with other species. These include *Formicoxenus nitidulus*, which nests within twigs in the nest mounds of *Formica rufa*. It is considered difficult to find and is best looked for on dull, warm days. *Solenopsis fugax* is often associated with nests of *Lasius* and *Formica* species, and it is thought to prey upon their brood (the young stages).

Some ants are very active and conspicuous, but a number of our species are small and unobtrusive and need to be very carefully looked for. Such searching, coupled with accurate identification, could yield useful information about the distribution and biology of these relatively little known British species. There is an active scheme for recording the distribution of ants in Britain (see p. 74).

In many ways, ants are easy animals to study. The main attribute needed is patience. Long-term observations can be made by revisiting nests repeatedly over a period of days, months or even years. Such long-term records are relatively rare as not many have had the persistence to continue with such work. A few hours spent by a nest at intervals throughout the season can yield valuable information about food intake, activity patterns and energy input. Daily and seasonal variations in these can be related to weather and other factors. These studies, coupled with observations under more controlled conditions in formicaria (p. 67), can give insights into factors controlling behaviour.

---

* References cited under the authors' names in the text appear in full in References and further reading on p. 75.

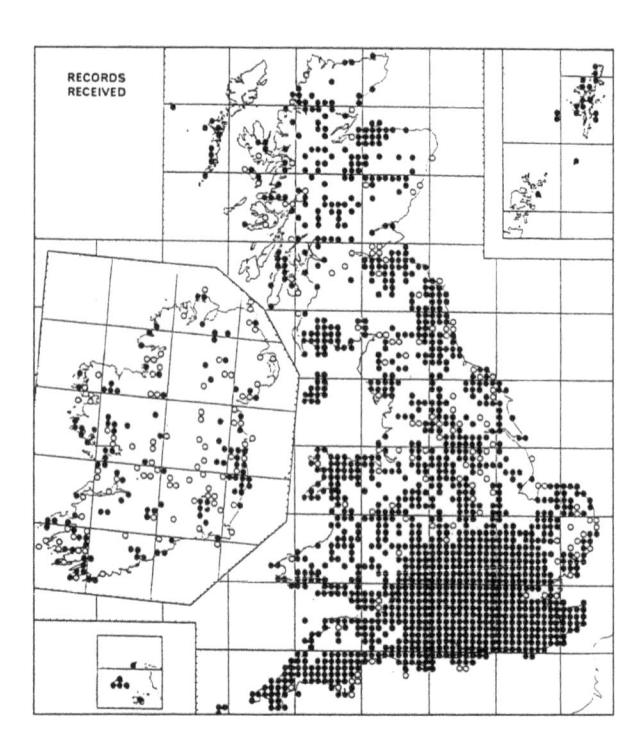

Fig. 3. The distribution of ant records received by the Ant Records Scheme before 1961 (circles) and between 1961 and 1979 (dots).

Because of their abundance, ants affect the environment in which they live in various ways. The environment within the nest can be highly controlled and ants, like man, are to some extent independent of prevailing external conditions. The availability of sophisticated, but quite inexpensive, data logging devices makes it possible to monitor variations in conditions in ants' nests throughout the day and the season, but valuable information can also be gained with simple equipment such as a mercury in glass thermometer.

Further afield, ants may affect populations of their prey, and therefore influence the growth of plants near nests. These effects can be revealed by comparing areas foraged by ants with ant-free areas. Prey populations can be monitored throughout the season, comparing situations where ants are present with others from which they are naturally absent, or have been excluded experimentally. Effects on prey may also be studied in laboratory colonies, where conditions may be controlled.

The evolutionary history of ants

The ants we see today have a long history. In 1967 a fossil ant called *Sphecomyrma freyi* was found in New Jersey, preserved in amber of Cretaceous age, possibly 80 million years old (fig. 4). These early ants were probably not common; only about 1% of the insects found in Cretaceous amber are ants. It was 65 million years ago that the ants

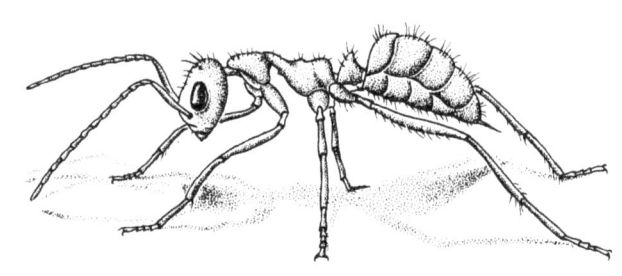

Fig. 4. The fossil ant *Sphecomyrma* (modified from Wilson, 1971).

began to move towards the dominant role they have now. Features of the anatomy of *Sphecomyrma* link it with modern ants on the one hand and solitary wasps on the other.

There is controversy about the origin of primitive ants, but most researchers regard them as close to the tiphiid wasps. A modern tiphiid thought to resemble ant ancestors is *Methocha fimbricornis* from the Philippines (fig. 5). A possible family tree for ants is given in fig. 6.

All ant species are social. Their social behaviour qualifies as the most advanced type, known as eusociality, because they exhibit three important features: there is a sterile worker caste, there is overlap of generations and females cooperate with their sisters in caring for the eggs, larvae and pupae, collectively called the brood. This type of true social organisation is found in the ants, a few species of wasps and bees, and the termites (Wilson, 1975).

The evolutionary origin of eusociality faces biologists with a difficult problem. How can evolution favour a sterile worker caste, if the workers never leave progeny of their own to carry their own genes into future generations? How can this apparently selfless behaviour arise? Darwin recognised this as a major problem with his own ideas; to him the presence of sterile workers represented 'one special difficulty, which at first appeared to me insupportable, and actually fatal to my whole theory' (Darwin, 1859). He suggested that such sterility could arise because these forms are useful to the community, and that selection might act at the family level. This, however, is only a partial solution and does not begin to explain how sterility could arise in the first place. In modern terms, the problem can be put as follows (Sudd & Franks, 1987).

'Why should a female choose to stay in the parental nest rather than attempt to start her own...family?...How could genes that code for helping the reproduction of relatives become more abundant in future gene pools, if such genes also led to an individual producing fewer of its own direct descendants?'

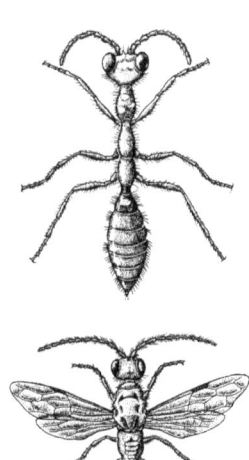

Fig. 5. A female (above) and a male (below) of the tiphiid wasp *Methocha fimbricornis* (after Williams, 1919).

Two solutions have been put forward: kin selection and parental manipulation (Sudd & Franks, 1987; Hölldobler & Wilson, 1990). It is beyond the scope of this book to explore these rather complex ideas in detail, but in essence the idea is as follows. Haldane (1955) pointed out that if one dived into

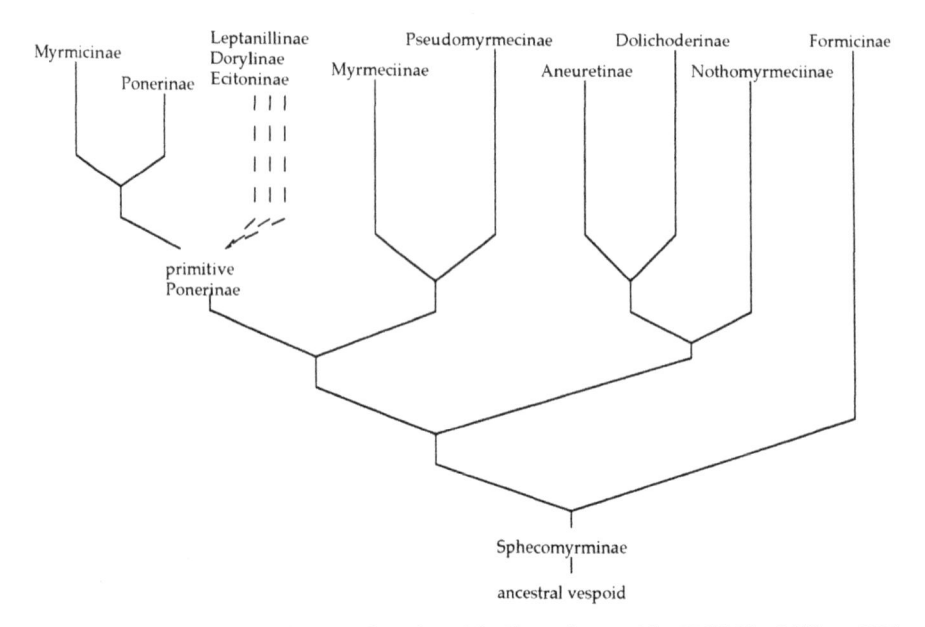

Fig. 6. An hypothetical family tree for ants (after Hölldobler & Wilson, 1990).

a stream to save a drowning child this would reduce one's own chance of passing on the 'helping others' (altruistic) genes because one might not survive. So how could such behaviour evolve?

A number of biologists (notably Haldane and R.A. Fisher) pointed out that if the child was related to his saviour then the risk to the rescuer's genes could be outweighed by the genes saved. This whole idea was formalised by Hamilton in a series of papers (1964, 1972) in which he put forward the idea of inclusive fitness, which takes into account the effect of doing things for relatives on an individual's fitness. Hamilton showed by calculation that natural selection will favour activities of an individual that benefit the relatives of that individual and will repress those that harm them. In this way social behaviour, and all that goes with it, would appear to have evolved (Berry, 1977).

2 Biology of ants

Nests

Almost all social insects build a nest of some sort. This gives them a home for the young and protection against attack, and against unfavourable weather conditions. The nest also provides a place to store food. In a few tropical species immobile workers act as food storage jars, filling their crop with honeydew and staying within the protection of the nest throughout their lives. Other species store seeds as food, and some tropical species establish fungus gardens in the nest.

Perhaps the most primitive nests are those simply excavated in the soil. Some species that build nests of this kind live mainly underground, feeding on root-sucking aphids and soil animals. Their nests are, typically, horizontal in layout, with chambers often following the roots on which aphids are sucking. *Lasius alienus* is a British species showing this pattern. Species that live above ground make vertical shafts with chambers. *Formica fusca* is one British species known to do this. Interesting observations have been made on nest architecture using plaster, latex or solder poured into the nest. This sets and provides a cast of the internal chamber and gallery structure. This technique should be applied only sparingly as it destroys the colony.

Some observations have been made on the behaviour of workers engaged in nest excavation, but we need to know more; this is an area where further contributions could be made, based on observations in the field and in formicaria. Most British species fall into this excavator category, and little is known about nest building in the less common species.

The other main type of nest found in British ants is the mound nest, which may be thatched or unthatched (Sudd, 1967). A number of species in the wood ant group of the genus *Formica* build thatched mounds. It is natural that the excavation process results in a pile of soil brought up from below ground during construction of the galleries and chambers. Some species carry this soil away from the nest, but others deposit it immediately outside the main point of entry into the soil, where it forms a mound. Unstructured mounds of this kind are most common in desert regions; in temperate areas definite, organised mounds with galleries are built.

We have only one true member of the unthatched mound group, *Lasius flavus*. *L. niger* and *F. fusca* sometimes build mounds, but these are never as extensive or permanent as those of *L. flavus*.

The wood ants use a wide variety of long, thin objects to construct mounds. Leaf stalks, pine needles and small twigs are amongst the materials used by British species. These mounds are not simply accumulations of such materials; they are carefully constructed and the ants undertake very extensive maintenance work on them. They seem to function in maintenance of a favourable environment for the colony. The internal environment of a mound nest is remarkably constant. Even ants with simpler nest architecture can regulate the conditions in which they live, and, more importantly, in which brood develops.

Ants are basically warmth loving animals. They are not active below 10° C and cannot produce young below 20° C. Thus, one of the most important things they need to do is to keep a warm nest.

There is a significant difference in nest function, and thus in siting and architecture, between tropical and temperate species. Whereas the temperate species nest under logs or stones, or build mounds, those in the tropics more often nest in trees or in rotting logs. For temperate ants rocks form good temperature regulators, warming up in the sun and releasing the heat slowly through the following hours. In spring a colony close to a rock can get on with such activities as egg-laying much earlier than a colony elsewhere in the soil.

Conditions are not uniform throughout the nest and workers have been shown to move the young stages around to keep them in the most favourable part of the nest at all times of day. This has been well worked out for a Japanese species, *F. japonica* (Kondoh, 1968). This is work that could profitably be repeated with some of the species in Britain.

Nest climate (or, more correctly, microclimate) is more closely regulated in those species that build mound nests than in other ants. As long ago as 1810 Huber suggested that the main function of these mounds was to regulate microclimate. That notion has been supported by extensive investigations since then. The most obvious climatic variable that might be regulated is nest temperature. The temperature 30 centimetres or so below the surface of a nest mound of thatch building *Formica* species varies very little (fig. 7). It is

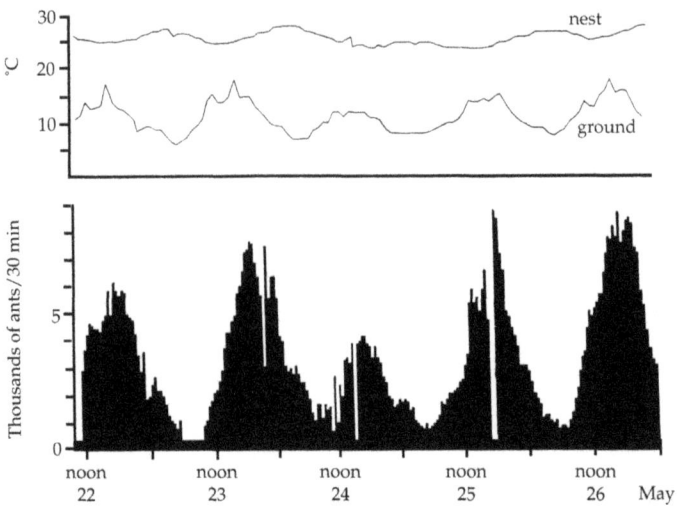

Fig. 7. Nest temperature and ant activity in a wood ant nest over a 4-day period (from Skinner, 1980a).

much more constant than that in the surrounding soil, and it is close to that which the ants prefer. How is the temperature regulated in this way? There seem to be two major factors involved: insulation and aspect. Insulation, slowing down heat loss, is achieved by thatching the outer crust of the mound, and maintaining the thatch in the face of rainfall erosion. Air trapped in the mound may also help to insulate the nest. Interception of the sun's radiation depends on aspect, as it does in a solar heating panel. The south-facing slope of the nest is shallower than the others, so that it intercepts the sun's rays at solar noon at right angles. The sun is probably not the only source of heat for the nest. The nest may be warmed by heat from the ants themselves along with that produced by the decay of the organic material, of which there is always much in the mound. There is some controversy over this latter point. Some researchers believe that decomposition makes a significant contribution to heat input whereas others consider that the heat produced by the ants themselves is much more important. Work on *F. polyctena* (not a British species) by Horstmann & Schmidt (1986) suggests that worker behaviour may have a very important part to play in regulation of temperature. Zahn (1958) goes as far as to suggest that the spring massing behaviour seen in *Formica* species actually functions to carry heat into the nest, within the bodies of workers which have been sunning themselves. This is not accepted by all researchers, but there is little doubt that the workers do contribute to the regulation of nest temperature less directly, by constantly repairing and redesigning the thatching of the mound. In several experiments in which nests were destroyed or disturbed, there was immediate and widespread worker activity to put things right. Clearly, the mound is not simply an unstructured accumulation of soil and fragments of vegetation.

There is much scope for further investigation in this area, particularly to see whether conclusions based on studies elsewhere apply also to British species or populations. There has been much confusion in the past about the taxonomy of European species of *Formica* and thus information from continental studies must be viewed with caution, especially if it is rather old. Much of this work would bear careful repetition and extension on critically identified species in Britain, now that the taxonomy of our species is clearer.

Ants can regulate the humidity as well as the temperature of the nest. Severe desiccation will kill all ants, but experiments show that some forms are more resistant than others. Ants placed in a dry container will eventually die, but tree-living species take much longer to do so than those from other habitats. In an experimentally created humidity gradient, workers have been shown to move the larvae and pupae into a preferred region where the humidity is similar to that found within the nest. For example, *Formica ulkei* selected a relative humidity of about 30% in this way (Scherba, 1959). The nest humidities which Scherba monitored were close to 30%, and the humidity range in the nest was lower than that in the surrounding soil. It is not clear how this regulation is achieved, but it would seem likely that nest structure is involved. In some species living in very dry habitats workers have been seen collecting water in various ways.

Ants in Britain do not encounter such severe problems. Only after very hot, dry spells does the soil in the upper part of a *Lasius flavus* mound become less than saturated with moisture. When this happens, the ants move downwards into cooler, more humid regions below.

Just as man's home is his castle, so is the nest of ants, and it may well have to be defended too. The defence often extends beyond the boundary of the nest itself into an area around it. This area, which is patrolled and defended, is the territory. Many animal species attack nests, including insectivorous birds such as the green woodpecker, *Picus viridis*, and man, but the most likely attackers are other ants of the same or different species. This risk is countered by a sophisticated recognition system. It has long been believed that each species, or even each colony, has a distinct odour. Only in recent times has it been possible to identify the complex mixture of chemicals involved. Because the quantities involved are so tiny, it was only with the advent of techniques such as gas-liquid chromatography that these odours could be characterised. The idea that ants produce species-specific odours seem to be well established but less is known about the basis of colony specificity that enables individuals to recognise friend or foe of the same species. Perhaps colony odour is derived from the environment and food of the colony, or perhaps a specific mix of volatile chemicals is inherited from the queen.

How efficient is this recognition? Invasion of a nest by a different species of ant elicits a violent and almost instantaneous attack from the occupants. If the intruder is of the same species, but a different colony, the attack may be just as violent, or it may simply involve offering less food to the newcomer, until it has acquired the colony odour. In most species, the response is so distinct that experimenters can use it to discover colony affinities of individuals. Workers of unknown affinity are placed on a test nest and the response of resident workers is compared with their response to 'control' workers from their own colony that are removed and then replaced. These observations give a clear answer to the question 'friend or foe?' This simple technique can be used to map territorial boundaries with some accuracy (fig. 8; Skinner, 1980a). The method is described further on p. 70.

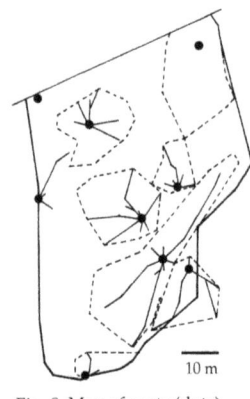

Fig. 8. Map of nests (dots), territorial boundaries (dashed lines) and trails of the wood ant *Formica rufa* in a wood (from Skinner, 1980a).

Caste and division of labour

A colony contains ants that may look very different from one another. At the simplest level, the colony contains males, which are generally short-lived and do no work, the queen, which founds the colony and lays eggs, and workers,

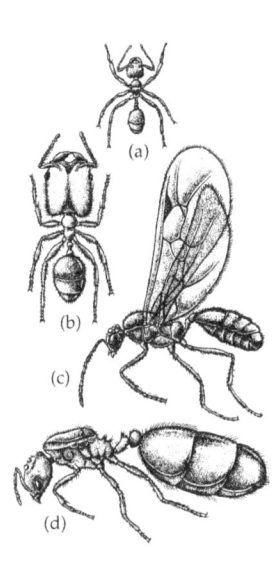

Fig. 9. Polymorphism in the non-British ant *Pheidole tepicana*, showing workers (a, b), a male (c) and a queen (d) (based on Wheeler, 1910).

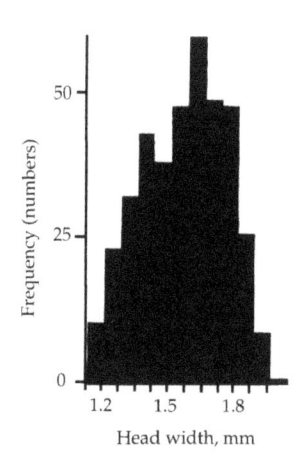

Fig. 10. Frequency histogram of head widths of *Formica rufa*.

which perform all the other tasks. Workers do not generally reproduce. Many workers lay eggs, but these are usually used as food. Less often, they give rise to males. In addition to males, queens and workers, various other forms are recognised. Ergatogynes are reproductive forms morphologically intermediate between queens and workers. In some cases these replace the queen. Such forms are found in the British species *Formicoxenus nitidulus* and *Leptothorax nylanderi*. Forms not known in any British species include the gamergates, reproductives with the morphology of workers, and dichtadiiform ergatogynes, queens with abnormally enlarged gasters.

In many species the worker caste may be further divided into sub-castes. The major (large) workers are called soldiers when their only job is fighting. Minor (small) workers are the best known forms, but media workers are also recognised in some species. This morphological segregation of workers into sub-castes is almost always directly related to a division of labour. These sub-castes are not found in any British ants, but that does not mean that the workers are all alike, in appearance or in behaviour. Size variation in the workers of British species is continuous, not discontinuous as in species with definite sub-castes. For example in the myrmicine *Pheidole tepicana* the major, media and minor workers are clearly distinguishable in terms of head width (fig. 9), whereas in the British *Formica rufa*, head width measurements show variation, but no obviously distinct forms (fig. 10).

Whether or not there are clear worker sub-castes, workers tend to change their jobs as they grow older. This phenomenon, know as age-specific polyethism, or temporal castes, is seen in all other social insects too, and is well documented in honey bees (Gould & Gould, 1988; Seeley, 1985).

For the first few days after emerging from the pupa, young workers generally remain in the nest, tending brood and the queen. Later they begin to make foraging trips from the nest, and become bolder in the investigation of foreign objects. Amongst British ants this change of job with age has been shown only in *Myrmica rubra* (Ehrhardt, 1931), *M. scabrinodis* (Weir, 1958a, b), *Leptothorax acervorum* (Buschinger, 1968), *Lasius niger* (Heyde, 1924), *Formica sanguinea* (Dobrzanska, 1959; Billen, 1984) and species of the wood ant groups (for example Otto, 1958; Rosengren, 1987). As far as we know, information does not exist for other British species. It would be useful to gather it for such species as *F. fusca/lemani*, other *Myrmica* species and *Lasius fuliginosus*. Such studies are probably most easily done on laboratory colonies (Chapter 4), but could also be done in the field. Some ants darken as they grow older so it is possible to estimate the age of individuals even if they have not been marked.

In a continental species, *Formica polyctena*, Dobrzanska (1959) investigated age polyethism by marking individuals with rings around the waist. Workers did not begin to forage until 45 days after emergence from the pupa. For the first few days they moved very little. At three days they began to take food from nest mates. At about six days they began tending brood. Then followed other jobs such as removing eggs, regurgitating food to young larvae and giving older ones bits of food. At about 27 days the workers began to show aggressive behaviour, and started spraying formic acid. Associated with these behavioural changes are physiological ones. Brood caring workers have well-developed ovaries, but the ovaries of foragers caught outside the nest have degenerated. Maxillary glands are large in nursing ants, but degenerate with age and are small in foragers.

Within these broad categories of tasks, called by German myrmecologists *innendienst* (inside work) and *aussendienst* (outside work), individual worker ants specialise in certain jobs. Again, this can be investigated by marking experiments. Otto (1958) showed that in wood ants the smaller outside workers tended to specialise in honeydew collection. This job specialisation has received little attention in the British species.

What determines caste?

In all the ants, bees and wasps, males arise from eggs that are unfertilized and therefore haploid (having only a single set of chromosomes). On the other hand females (queens and workers) arise from fertilised eggs which are diploid (having two sets of chromosomes). Amongst females, caste is determined by physiology and not by genetics. That is, all fertilised eggs have the potential to become either workers or queens; whether a particular fertilised egg develops into a worker or a queen does not depend in any way on its genetic make-up. A number of other factors seem to be important. The presence of the queen is one such factor. In colonies from which queens were experimentally removed after eggs had been laid, more larvae developed into queens than would otherwise have done (Brian & Carr, 1960).

In *Formica polyctena* the proximity of queens has been shown to be crucial. Queens of this species lay their early-season eggs in the top of the nest, and then move down into the lower chambers. These early-season eggs have a strong tendency to develop into new queens. In experiments in which existing queens were prevented from moving away, these eggs failed to develop into new queens, but became workers instead. The situation is similar in a related species, *F. pratensis*, but in this case the queen-suppressing effect of the presence of an existing queen can be overridden by a very large worker:queen ratio (600:1) (Gösswald & Bier, 1953). The presence of a queen also seems to inhibit worker reproduction. In some species more eggs were laid in colonies without queens than in colonies with queens. For example, 50 *Leptothorax unifasciatus* laid 140 eggs when queens were absent, but only 60 in the presence of queens (Bier, 1954).

It is not clear how queens exert this influence over brood development. In honey bees it is well known that the queen exerts a similar influence via a chemical message or pheromone called queen substance, and this has been identified as E-9-oxo-2-decenoic acid. Does such a controlling pheromone exist in ants? The main evidence that it does comes from work with *Myrmica* species in Britain. Dead queens, if regularly replaced, inhibited the growth of larvae (Carr, 1962), and later it was shown that this inhibiting effect was due to a fatty acid that could be extracted from the heads of queens. More recent work has implicated other factors as well. Firstly, cytoplasmic factors are involved. Eggs do vary in their ability to become queens or workers, perhaps because of differing amounts of reserves in the cytoplasm. Small, quickly-laid eggs tend to become workers, and slowly-laid eggs, large or small, can become either queens or workers. Secondly, young queens tend to lay worker-biased eggs, and older queens lay more queen-biased eggs. A third factor of importance seems to be temperature. If this is lower during egg formation than it is after the egg has been laid, the egg will tend to develop into a worker.

The path of development, to worker or queen, remains undetermined and susceptible to environmental modification through much of the larval and pupal stage. The quality of the food supply is important. A larva in a well-established colony will grow well and hibernate over the winter. If the food supply is still good in spring, the larva will continue towards queenhood, but if not, it will develop quickly into a worker. Queens influence the path of brood development through their effect on worker behaviour. In the absence of queens, workers give extra food to the larger larvae, those most likely to become queens. When queens are present, larger larvae receive less food, and both workers and queens may attack them, either killing them or inducing early metamorphosis into workers. Temperature is important too. If it exceeds about 20 °C in the last larval stage, workers tend to develop. Since queens do not develop at high temperatures Brian (1977) suggested that *Myrmica* is adapted to, or perhaps even a prisoner in, cold climates.

Hölldobler & Wilson (1990) summarised this complex picture very nicely:

'...six factors...determine whether a *Myrmica* female will become a worker or a queen: larval nutrition, winter chilling, posthibernation temperature, queen influence, egg size, and queen age. The next question should logically be, what is their relative importance in nature? The clearest way to view the entire caste-determining system is to regard it, metaphorically, as a series of checkpoints arranged more or less in sequence. An egg "aspires" to develop into an adult queen. This ambition is "approved" by the colony, providing the following two checkpoints are passed. First, has the larva been through diapause and chilled to resume full development? Second, has the larva reached the requisite size by the onset of adult development in the final larval instar? In addition, are the mother queens nearby and potent, and is the colony young? If so, borderline cases are more likely to fail the queen test and be consigned to workerhood. Taken together, the caste-biasing factors make it more likely that the *Myrmica* colony will produce new queens in the spring and also when it is large and robust – the conditions under which it can most profitably invest in reproduction.'

This complex situation has been worked out in species of the genus *Myrmica*, but what of other genera? The other group in which very detailed studies have been carried out is the *Formica rufa* group, and this shows a number of differences from *Myrmica*.

The first major difference is that the ants hibernate without any brood. Queens move to the upper parts of the nest when the temperature rises in the spring, and begin laying eggs. At first the temperature is too low for fertilisation to occur and so these very early 'winter eggs' become males. When the temperature rises above about 19 °C, the next batch of winter eggs laid are fertilised and can become workers or queens. *Formica* can regulate nest temperature, but small colonies do this less effectively than larger ones, and thus tend to produce more males. The eggs laid in summer all become workers.

The final destiny of a later, winter egg, worker or queen, seems to be determined in the first three days of larval life. Again, the presence of queens is important. When none are present these winter eggs always develop into queens. As we have seen, in a natural colony the larvae are effectively queen-free at this stage, because the queens have gone deep into the nest.

Hibernated workers are very active and have well-developed food glands, and this increases the tendency of winter eggs to develop into queens.

Thus, the system is similar in some respects to that found in *Myrmica*. Eggs are biased towards worker or queen; there is a seasonal trend in this bias; queens suppress queen formation; workers vary with season in their ability to nurse; and food availability and temperature are important.

Most other studies have shown only minor variations on these basic patterns, but there are some very different situations, and further study of other species would be worthwhile.

Communities of ants

If one species of ant is found in an area several other species will be found co-existing with it and the range will be predictable. Such an assemblage of species is called a community. Elton (1966) has analysed the formation and maintenance of animal communities. In a new habitat, some species cannot gain admission because they have not got the dispersal power to reach the habitat. Of those that do colonise, some species will not survive for very long, and others will become established. This process of elimination will produce a relatively stable assemblage.

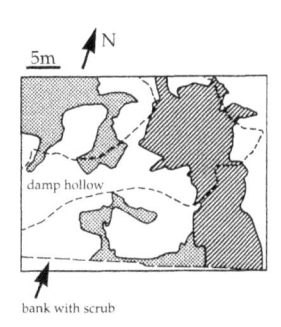

Fig. 11. An ant community on a Dorset heath. *Lasius niger* (hatched) lives in the shade of a bank and in a damp hollow (dashed line), *L. alienus* (stippled) on higher ground, and *Tetramorium caespitum* (white) in between (from Brian, 1983).

In Britain, the most extensive studies of ant communities are those of M.V. Brian and his associates at the Furzebrook Research Station in Dorset. The following account, which relies heavily on Brian's work, gives a couple of examples. More details can be found in his original papers and his excellent book (Brian, 1977).

In a patch of lowland heath, Brian found that 10% of the total area was covered by mounds of *Lasius flavus*, with a nest density as high as 1500 nests per hectare. In the most densely populated south-facing regions the nests were evenly spaced. This apparent mutual avoidance can be taken as evidence for intra-specific competition. Presumably, a queen attempting to found a new nest within the territory of an existing nest would be killed.

In some places, particularly on the valley floor, *Myrmica scabrinodis* and *M. rubra* replaced *Lasius flavus*. Here it was much cooler, and the drainage was not so good as it was on the valley sides, because of the nature of the sub-soil. Another community was made up of *Lasius flavus* and *L. niger*. The *L. niger* were found under stones, interspersed amongst the *L. flavus* nests. There was some evidence for direct competition; *L. niger* workers were often found with the mandibles of a detached *L. flavus* worker head clamped around one leg. Less direct evidence of interaction between the two species comes from the finding that queen production in *L. flavus* was much reduced in places where *L. niger* was present.

On limestone soils in the south, Brian found *L. flavus* and *Myrmica rubra* together. Here the *Myrmica* could nest only in the shady areas and was absent from sunnier places. The two species were separated in their foraging too. *Myrmica* found food above ground, whereas *L. flavus* fed below ground level.

Dry, base-poor soils in the south support a plant community dominated by Ling, *Calluna vulgaris*. A very different ant community is found here. In the best areas, *Tetramorium caespitum* was found with *Lasius alienus* (fig. 11). *Lasius niger* was frequent too, and the large black *Formica fusca* was present in smaller numbers. In addition there were occasional nests of *Tapinoma erraticum, Formica cunicularia, F. candida, Myrmica ruginodis, M. scabrinodis* and *M. sabuleti*. *Lasius flavus* occurred in one or two places. The parasites *Strongylognathus testaceus* and *Anergates atratulus* were also occasionally present in nests of the host, *Tetramorium*.

Reproduction

All ants have the same basic life cycle. Eggs are laid only by queens. They hatch to give a legless larva which is fed by workers. It moults, usually three times, and finally pupates. In this inactive pupa (in some species encased in a silken cocoon and in others not) the larval tissues are completely reorganised to form the adult (imago). This will emerge from the cocoon in a pale, callow state. Recently emerged workers are recognised by their pale colour, but they darken within a few hours. Plate 4 shows various immature stages of British ants.

As indicated elsewhere, in the ants the specialised females, called queens, are generally the only ones that lay eggs. Even if workers do lay eggs, these are nearly always trophic eggs, used as food. Occasionally, worker eggs hatch producing males, which in the Hymenoptera come from unfertilised (haploid) eggs (p. 12). Fertilised eggs, with the potential to become new queens or, more often, workers, can be laid only by queens, because only queens mate. Colonies produce the sexual forms (that is, winged, unmated females, and males) at various times of year, depending on species. In the wood ants, sexual forms appear in May, having been reared on food stored over the winter. In most other species sexual forms appear in midsummer.

Generally, all the nests in a wide area will produce their winged forms on the same day, which is usually a warm and sultry one. Millions of these 'flying ants' may emerge on such a day. Most will not live to reproduce, as they will be eaten by insectivorous birds and by other insects.

There is much still to be discovered about the pattern of nuptial flights. Most seem to fall into one of two broad types, which Hölldobler & Wilson (1990) refer to as the female calling syndrome and the male aggregation syndrome. In the former, wingless females remain on the ground near the nest, releasing sex pheromones, 'calling' winged males to them. This pattern is generally found in more primitive species with small colonies and a low rate of production of sexual forms. In the male aggregation syndrome, the males congregate in various prominent places such as the tops of hills, trees or buildings in large swarms, made up of males from many colonies. The females fly into these swarms, where they presumably mate, and then disperse, eventually to shed their wings and begin founding a new nest, if they survive long enough.

A successfully inseminated queen sheds her wings when these have served their function of dispersal. The muscles that operate them are to be used as a source of food in the difficult days ahead. In many species the queen alone must feed the first batch of larvae that hatch from her first-laid eggs. Material derived from the breakdown of the wing muscles contributes to the food for the larvae as well as to the eggs themselves.

In some ant species the queen builds a small nest and leaves it periodically to find food for her brood. This 'partially claustral' pattern is thought to be the primitive state. In more advanced 'claustral' forms the queen builds a small nest chamber and seals herself into it to live on stores until the first workers emerge.

Some ants found colonies rather differently. In the wood ant *Formica rufa*, for example, the usual pattern of colony founding is for mated queens to return to the home nest and then possibly move to a new site with some workers.

A more interesting pattern emerges when mated queens of, for example, *F. lugubris* fail to get back to their own nest. They may then invade the nest of a different species, in this case *F. fusca*. When this happens, the *F. lugubris* queen eliminates the *F. fusca* queen and takes on the role of queen to the *F. fusca* workers. She also lays eggs. These, and the larvae that emerge from them, will be tended by the *F. fusca*. Eventually all the *F. fusca* die off, and because they are not replaced, they leave a pure *F. lugubris* colony. If this were a common way of founding, one would expect to find mixed *F. lugubris/fusca* nests regularly, but this is not the case. This method, known as temporary social parasitism, is therefore probably less common than the budding described above.

F. rufa is a more regular temporary social parasite, and for this reason most of the colonies are single-queened (monogynous). The hosts are *F. fusca* and *F. lemani*. The approach of the *F. rufa* queen is not very subtle; she simply barges into the nest. She often perishes here at the hands of the workers of the other species, but enough attempts are successful to make this a fairly common species, although it is currently declining due to loss of habitat in Britain.

The rare *F. exsecta* founds its colonies in a similar way but the small shiny *F. exsecta* queens have a more subtle approach to colony penetration. Often, they are carried into the nest by workers or enter in a stealthy way, unlike the more direct approach of *F. rufa*. They do not elicit such hostile behaviour as other queens.

Colony founding is similar in *Lasius fuliginosus* and *L. umbratus*, and other members of these two species groups. The host is any member of the *L. niger* group of species. In this case, however, the relationship is obligatory; the parasite species have no alternative method of colony founding. Hölldobler (1953) showed that a queen of *L. umbratus* first kills a worker and runs around with it in her mandibles before attempting nest entry. The host queen is disposed of. It is not known how this is achieved, but in the Austrian *L. reginae* she is rolled over and throttled.

In the non-parasitic species that found colonies 'from scratch', the first workers to emerge are rather small and timid compared with those found in a mature colony. These features seem to be adaptive in the sense that there is a balance between the number of workers and their size. The queen has limited resources and she can use these to produce a few large workers, or many smaller ones. A large number of workers will be able to do all the jobs of food finding, nest enlargement and so on, but if the workers are too small they will not be able to exploit the normal food and nest sites of the species. It should therefore be possible to predict the optimum number of newly emerged workers and this predicted number of workers has been shown to be close to true number in at least one species, *Solenopsis wagneri*.

After a hazardous early stage the colony, if it survives, will enter a period of sustained exponential growth. In those species that have been studied, the growth curves show a typical sigmoid form, but it is also clear that the factors controlling the growth of the population are much more complex in ants than they are in the non-social insect species.

Eventually, before it reaches its maximum worker population, a colony begins to reproduce. Much work has been done to identify the factors controlling the onset of this reproductive phase in the colony's history in the genus *Myrmica* by Brian and his team at Furzebrook. In one series of observations Brian was able to show that colonies often produce males before any new queens appear, and he described a colony in this male-producing phase as 'adolescent', as opposed to the earlier 'juvenile' phase and the later 'mature' phase in which queens are produced as well. Brian also showed that the colonies enter the adolescent and mature phases due to a dilution of queen influence. In *M. ruginodis* the switch occurs when there are 900 to 3000 workers, but the crucial figure is the ratio of workers to queens, and in this species the relevant ratio is 1000 workers per queen. It has been shown that the queen dilution effect arises when she no longer touches each of the workers often enough to have proper control over their behaviour towards the brood. This influence seems to act via a chemical, because dead queens are effective when intact but not when their lipids have been extracted. Brian's summary of the situation is illustrated in table 2.

Table 2. *A summary of influences that determine caste in ants (after Brian, 1977)*

	Behavioural type	
	Queens present	Queens absent
1. Worker-biased larvae	Nurses feed these actively in preference to other larvae and they metamorphose into workers	These larvae are neglected and they stop growing part way through the third instar
2. Labile female larvae	Workers (mainly foragers) attack these larvae during the stage in spring when they are signalling their ability to become gynes by means of a cuticular secretion	Both workers and foragers feed these larvae on solid protein-rich foods, focusing on a few at a time
3. Male larvae	Neglected but not attacked	Fed in preference to worker-biased larvae but not labile female larvae that are forming gynes
4. Egg formation and laying	As long as the queen is laying or there is a cluster of reproductive eggs of either sex, young nurse workers lay imperfect eggs that are useful only as food (trophic eggs). Very young workers (<3 weeks old) will lay reproductive eggs that form before they become sensitive to queens	Young workers lay large perfect reproductive eggs that are unfertilised but capable of developing into males parthenogenetically

In other species it has been shown that workers bite repeatedly at the larvae, and this causes the larvae to develop into workers.

Apart from this work on *Myrmica* species, little is known about the timing of the change from juvenile to adolescent and mature stages. Thus it is not known whether the same pattern is found in other species. Nonacs (1986) reported that some colonies will produce only males in a particular season, whilst others produce only queens and yet others a mixture. It is not known whether the strategy of an individual colony changes from year to year.

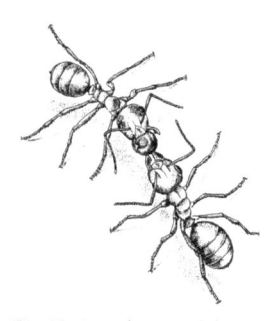

Fig. 12. A worker ant giving food to another (after Hölldobler in Dumpert, 1978).

Feeding

Ants spend much of their time collecting food. Worker ants are exclusively responsible for this job within the colony. Males rarely feed, and the queens and the young stages are provided with food by the workers. Most other animals collect food for themselves alone, or possibly for their own young, but much of the food collected by an individual of a social species is not for itself at all (fig. 12).

A consequence of this is that food has to be carried back to the brood and queens in the nest. It is therefore possible to monitor the food intake of ant colonies. In most other invertebrate animals, specialised techniques are required to monitor food intake, but with ants much can be learnt from simple observational studies. In Britain most of this work has been done on species that form obvious trails, especially the wood ants (*Formica* species) and one or two others such as *Lasius fuliginosus*. Such studies reveal that ants subsist on a wide range of food items. *Formica rufa* have been shown to carry many types of prey (table 3). Some others are shown in table 4. A list like this can be produced quite quickly by observing trails leading to the nest, but it may take rather longer to get an idea of the quantities of material brought back (fig. 13).

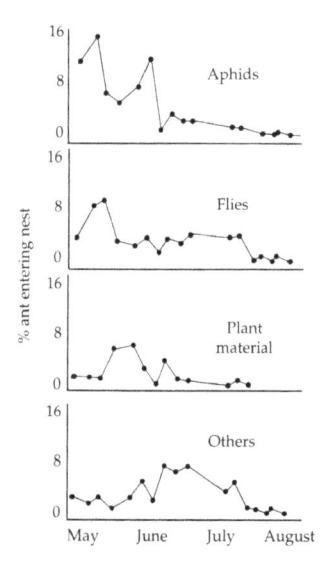

Fig. 13. Food income to a nest of Formica rufa (from Skinner, 1980b)

Table 3. *Food items recorded being brought back to a* Formica rufa *nest in northwest England*

---
aphids
bark lice (Psocoptera)
beetles
earthworms (usually in bits)
bumble bees
a variety of small flies, especially St Mark's Fly (*Bibio* species)
woodlice
moth caterpillars (especially winter moth *Operophtera brumata*)
small spiders

In the study of wood ants described above, five-minute periods of observation were long enough to give useful information, but longer observation periods would be needed if there were less traffic of food items along the trail. To quantify food intake of a colony, it may be necessary to make observations over periods of an hour or more.

The importance of each type of food for the colony is not shown by numbers alone. For example, a caterpillar brought back by wood ants may be many times larger than a single aphid. A single observer can look at only one trail at a time. Although it may be possible to get some idea of what is being carried back by simply looking, very often it is necessary to confiscate the food item for closer inspection. This inevitably disturbs the ants. The deprived worker will release chemical alarm signals which affect other foragers on the trail. It is rarely possible to study an animal without such an 'observer effect', but this should be minimised wherever possible. One way to reduce disturbance is to install permanent sampling devices on trails (technique p. 72). These do disturb the ants, but the ants seem to get used to them. Automatic devices make longer-term sampling possible and remove the 'one-observer-one-trail' constraint. The sampling limit depends only on the number of devices available, although sorting the collected material can be very time-consuming (Skinner, 1980*a*).

Ants are basically omnivorous animals. All but the most specialised include both animal and vegetable components in the diet. Ants do not generally eat solid plant material directly, and must suck up juices. The fungus growing ants collect leaves and then grow fungi on them, and then eat parts of the fungus. None of these ants occur in Britain, where the vegetable component of the ant diet consists almost exclusively of honeydew, a sugary fluid excreted by aphids. Ants have sometimes been seen chewing at the bases of bud scales or drinking nectar in flowers, but these habits do not seem to be widespread. Indeed, some flowers seem to have ways of keeping ants away from the nectar; ants are not good pollinators. The situation is different in the extrafloral nectaries on the fronds of bracken fern, *Pteridium aquilinum,* and at the base of the leaf stalk of broad bean or common vetch. Ants visit these structures to lick the sugary nectar, and the plants evidently derive some benefit because the ants protect them from herbivores. This association is commonly described as a mutualism, but it does not always result in overall benefit for both parties; the balance of costs and benefits for the ant and the plant often depends on local circumstances (Koptur & Lawton, 1988; Rashbrook and others, 1991).

By far the commonest plant-derived food for ants is honeydew from plant-sucking bugs. Phloem sap is rich in sugars, particularly the disaccharide sucrose, but relatively poor in amino acids. Plant-suckers, notably aphids, require a higher ratio of amino acids to sugars than the phloem provides. They therefore take in much more sugar than they need, and the excess sugar and water is voided from the

Fig. 14. A wood ant worker 'milking' an aphid (after Dumpert, 1978).

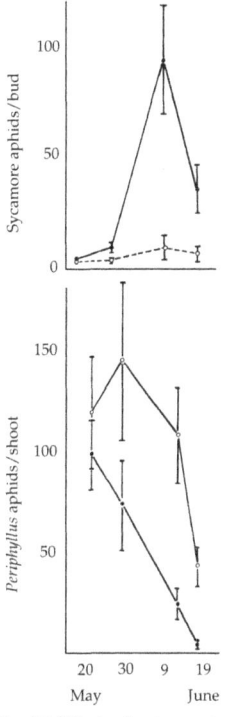

Fig. 15. Effects of ants on a tended aphid (*Periphyllus testudinaceus*) and an untended aphid (the sycamore aphid) on sycamore. Ants had free access to some trees (open circles) and were excluded from others by sticky-banding (black dots) (from Skinner & Whittaker, 1981).

anus as honeydew (fig. 14). Ants avidly tend aphids to get this material. A very wide range of such ant/aphid mutualisms has been observed. Some of the major conclusions and unanswered questions are summarised below.

All aphids produce honeydew, but not all species of aphids are tended by ants. This was shown for the wood ant *Formica rufa* and the aphids on sycamore trees in a northern English woodland. The two aphid species common on sycamore leaves were *Periphyllus testudinaceus*, which ants tended, and the sycamore aphid *Drepanosiphum platanoidis* which was not tended. Tending affects aphid numbers. Fig. 15 shows how the numbers of the two species changed over a season, when ants were kept away from them by sticky bands around the stems. The tended species benefited enormously from the presence of the ants, whereas the non-tended species did not. Indeed, the non-tended species actually suffered from the ant attention. The reason for this was easily found by looking at the prey items that the ants were carrying back along trails to the nest. Many specimens of the sycamore aphid were taken as prey.

Questions still remain. Why do ants attend one species but not the other? Is it because *P. testudinaceus* lives in colonies, so that foraging ants can 'milk' honeydew from several at a time, whereas individuals of *D. platanoidis* are more widely scattered? And how does ant attendance enable aphids to do better than they would if the ants were not there?

If honeydew were not removed by the ants it would accumulate around the aphids causing physical problems and enhancing the growth of mould, which could be detrimental to the aphids and the plants that support them. In the non-tended species, and in other species which are sometimes tended and sometimes not, droplets of honeydew are flicked away from the body with the hind legs or the tip of the abdomen, or by contraction of the abdomen or rectum. Some ants, such as members of the genus *Leptothorax*, will sometimes lick up such honeydew from the ground or from leaf surfaces. Others, such as the tropical *Acropyga*, are totally dependent on honeydew (Sudd, 1967). In Britain, *Lasius* species are probably the group most dependent on honeydew, although species of *Formica* and *Myrmica* take a considerable amount.

Many species of aphid have special adaptations for ant attendance. Some, such as *Forda formicaria*, are found only in association with ants, and others do not thrive in their absence. For example the sycamore-dwelling species *Periphyllus testudinaceus* is never very common except when attended by ants.

The most obvious adaptation to ant attendance involves the cornicles. These are paired structures on the aphid's abdomen, used in defence. When an aphid is approached by a potential predator a waxy substance is squirted from the cornicles, and this usually deters further attack. In many (but not all) frequently-tended species these cornicles are very small. Some aphids will also respond to potential predators by kicking, walking away or even dropping from the plant (Rotheray, 1989, 1994). In species that allow ant attendance, these behaviours seem to be suppressed. A further adaptation involves the so-called 'trophobiotic organ', a basket of bristles around the aphid's anus which serves to hold the honeydew droplet for the ants.

As we have seen, ant attendance increases population numbers of aphids such as *P. testudinaceus*. In colonies of aphids of the genus *Aphis*, ant attendance reduces the tendency to produce winged forms (alates). This, in turn, allows the colony to grow larger by preventing dispersal. Some research workers believe that this effect may be due to the presence, in the secretions of the ants, of juvenile hormone (JH) mimics. JH promotes winglessness, and similar molecules in ant secretions may have the same effect (Kleinjan & Mittler, 1975). This has been noted in only a restricted range of ant/aphid associations. In other situations, ants may nibble the wings of alates, preventing them from flying away (Kunkel, 1973).

The most obvious way in which ants increase aphid populations under their care is by removing, or deterring attack by, the aphid's natural enemies. This effect can usually be shown very clearly by simply comparing the numbers of such enemies in tended and non-tended situations. Although it is natural to suppose that the aggressive behaviour of the ants leads to a reduction in the numbers of aphid predators and parasites, not much quantitative work has been done on this aspect of ant/aphid mutualisms. Here is a rich field for research, both in the wild and in laboratory colonies of aphids on potted plants. A productive approach to such studies is illustrated by some of the work that has been done by, for example, Way (1954) or Bristow (1984). Way (1954) used several experimental techniques. The attendant ant, in this case *Oecophylla longinoda*, was excluded from colonies of plant-sucking bugs, in this case a scale insect *Saissetia zanzibarensis*, either by completely removing the ants or by ringing the plant stem with bands of sticky grease which the ants could not pass. Some of Way's experiments involved potted plants supporting the scale insect and the ant.

Another way in which ants appear to aid aphids is by removing the potentially harmful honeydew. In the association that he studied Way showed that the scale insect did much better when attended by ants, but that the benefits of attendance could be partially mimicked by washing away the accumulated honeydew with water.

A further component of the ant effect involves the rate of reproduction of the aphids, which in the tended species seems to be increased in the presence of ants (for example, El-Ziady & Kennedy, 1956).

In general, little is known about the behaviour of ants and aphids except for qualitative descriptions, but Douglas & Sudd (1980) give a very detailed account of the relation between the ant *Formica lugubris* and the aphid *Symydobius oblongus* on birch. For example careful observations of ant and aphid showed that although the aphid was tended for only 14% of the time, it released 84% of its honeydew during this period. A touch of the aphid's abdomen by the ant led almost invariably to the production of a droplet; the idea that ants 'milk' aphids was thus appropriate in this case. This study also showed that the aphid could communicate with the ant, signalling its readiness to release a droplet by raising the abdomen. In its partnership with the ant *F. polyctena*, the aphid *Lachnus roboris* conveys the same message by raising its back legs. Similar careful studies on other ant/aphid partnerships are likely to yield useful information. Ants are normally very aggressive and it is still not clearly understood how this behaviour is suppressed in their relationship with aphids. Kloft (1959) suggested that the rear of an aphid's abdomen resembles an ant worker offering food (fig. 16). The ant might mistake the combination of features on the aphid's rear for the front end of a sister worker, and switch into a behaviour pattern resembling the normal food-sharing behaviour that occurs between sister workers. Kloft suggested that the normal leg waving or kicking that most aphids perform when disturbed has become ritualised as an interspecific signal. On the other hand, many non-aphid Homoptera such as scale insects and mealy bugs, which also produce honeydew, are totally unlike ants in appearance, but are just as avidly attended. Perhaps the resemblance noted by Kloft is simply coincidental. The whole idea needs to be tested by careful observation and experiment.

Fig. 16. Ant's eye views of the abdomen of an aphid (above), with hind legs, and the head of an ant offering food (below), with antennae (after Kloft, 1959).

A further way in which aggressiveness may be suppressed is by chemical communication. Some very closely ant-associated aphids are known to produce substances other than honeydew which are attractive to ants (Hölldobler & Wilson, 1990). Also it is possible that ants scent mark aphid colonies. Both these ideas need further investigation.

Apart from some very unusual examples, all ants need some animal food in the diet. Again, as with honeydew, species differ in the extent to which they depend on such foods. In the tropics army ants rely almost entirely on animal food, whereas leaf cutter ants eat very little, if any. In Britain, all our species are omnivorous to some degree. Little is known about exactly how ants find and catch their prey. It is clear that wood ants can see from a distance of 10 cm or so. Some ants may be able to detect vibrations caused by movement of the prey. All ants have a good sense of smell which may be important in prey detection. Small invertebrates are caught by a slow approach followed by a final pounce. In those ant species with stings, poison is then injected. In others, the prey is sprayed with poisons from glands. Fellow workers nearby often help to kill the prey. However, many of the prey species seen being carried along ant trails actually have very effective escape mechanisms. This has led to the suggestion that much prey is caught only when incapacitated in some way. Some research workers have suggested that many ant species are acting mainly as scavengers when they bring back such food. This may well be the case for wood ants. There is much scope for further work on the food income of nests, both in terms of what is brought back and in terms of energy flow.

We know little about the differences in diet between different species or different colonies, and changes of diet with season. Only recently has there been any real appreciation of the food of the non-trail-forming *Lasius flavus*, despite the fact that it one of our commonest species. A summary of some of the conclusions reached for the British fauna is given in table 4. Much information could be added by careful studies of other species. The data available for *Myrmica* are very generalised, but other observations led Brian (1977) to expect differences within the genus. For instance, *M. scabrinodis* is more of a hunter in short vegetation than its close relative, *M. ruginodis*. Detailed inventories of food have not been made and could be constructed by careful observational work. This requires little equipment, but plenty of patience and time.

Table 4. *The prey of British ants (excluding* Formica *species, see table 3) (based on Brian, 1977)*

Myrmica	aphids, springtails, larval and adult flies, spiders, others
Lasius flavus	soft-bodied mites, beetle larvae (mainly wireworms of two species), woodlice, *L. flavus* workers and queens
Lasius alienus	centipedes such as *Geophilus*, wireworms

In addition to animal prey and honeydew, some ants eat seeds and even collect and store them for later use. This is a very common phenomenon in ants, especially in desert environments, where food, in the form of the seeds of ephemeral desert plants, may only be available for a short time. In Britain, *Myrmica* are known to collect seeds of tormentil, *Potentilla erecta*, but it is not known whether or not these are eaten. In the spring *Tetramorium caespitum* collects a wide range of seeds and feeds them to larvae, after chewing them (Brian, 1977). In other cases of seed collection noted in British ants, the seed has an oily body, the caruncle, which the ants consume, leaving the rest of the seed undamaged and thus able to germinate. In these cases the ants are performing a dispersal function for the plants. For example, seeds of gorse, borage, viola and primula are known to be treated in this way.

Food provides not only materials for growth, development and reproduction, but also energy for all activities. Attempts to look at energy budgets of ants' nests are rare. The work of Nielsen (1972) on *Lasius alienus* in Denmark can serve as a model and starting point for anyone interested in doing such work on other species.

Foraging strategies

The vast number of known ant societies use a correspondingly diverse array of foraging strategies to locate and secure their food. According to Hölldobler & Wilson (1990), the behaviour we call foraging can be split into three main phases or categories: hunting, retrieving and defence. The more primitive ant species tend to have workers which forage alone, carrying or dragging the food to the nest without help. Such workers seem to navigate with the aid of light patterns, gravity, and possibly also chemical cues. Wood ants seem to use the disposition and shapes of trees and shrubs against the sky, together with the sun's direction. On the other hand some species, notably *Lasius fuliginosus* in Britain, rely more heavily on chemical trails. Many species make long-lasting trails, which they follow from year to year. Memory of foraging sites seems to be good. There is evidence that each worker is responsible for only a restricted part of the territory, so that the area is partitioned amongst the worker population. Thus, each worker has only a small area to remember.

There are a number of ways in which workers recruit help to exploit a newly discovered food source. A worker arriving back at the nest with food from such a source may simply alert others to forage. They then rush out at random, but there is a high chance that some will find the food. A more advanced state is seen when workers follow a successful worker to a new source by 'tandem running'. At a more advanced level are species such as *Myrmica* species, *Tetramorium caespitum, Lasius niger* and *L. fuliginosus* which are known to lay scent trails. Little is known about the trails in these species. For example, are they directional? We know nothing at all about the situation in most of the other British species, although there is some suggestion that the parasitic ant *Formicoxenus nitidulus* (which occurs in Britain) follows trails left by its wood ant host (in this case the species *Formica polyctena*, not found in Britain) (Elgert & Rosengren, 1977). An intriguing question arises as to what *F. nitidulus* does when it is in the nests of non-trail-laying species such as *F. rufa*. Much useful work could be done on ant trails, on both wild and laboratory colonies, using simple but effective techniques. In interruption experiments, the area in front of the ant is rubbed, for example with a finger, to interrupt any chemical trail. Ants reaching the rubbed place stop short and there is a delay before they cross it. Similar results can be obtained by rubbing with other implements or by putting Sellotape over part of the trail. In trail displacement experiments, a trail is established over a sheet of paper and this is removed and replaced, as it is or after cleaning, back to front or with a fresh one. This experimental technique could form the basis of some interesting work, especially on laboratory colonies. Feynmann (1985) gives some further details.

The trail can be made visible by dusting talc or lycopodium powder onto the area where it is thought to have been laid, and then blowing away any powder which is not stuck down. This works well with *Lasius fuliginosus*.

An artificial trail can be made by dragging the abdominal tip of a dead ant along a line to see whether or not live workers will then follow this trail. If they do, further experiments could explore the efficiency of different glands in producing a followable trail. The glands would need to be separated, by careful dissection, for such experiments. Some ants will follow trails laid by another species. In interspecific trail following experiments, different species are tested for their ability to follow trails formed by others. Hölldobler & Wilson (1990) list ants known to be trail formers and followers, together with the source of the trail substance, and references.

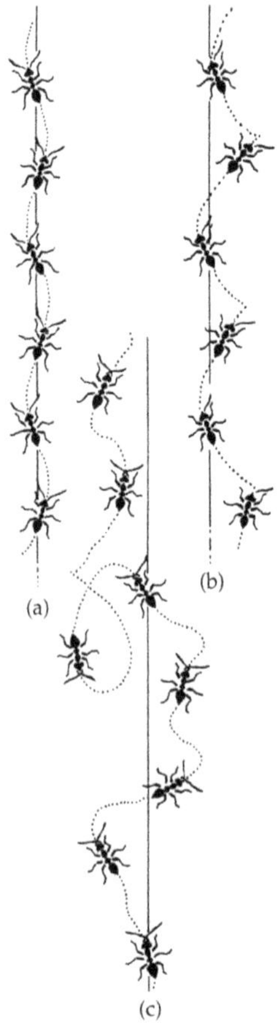

Hangartner (1967) has shown how *Lasius fuliginosus* follows a trail (fig. 17). These ants do not follow a chemical deposited on the ground but walk along the 'tunnel' formed by its vapour, waving their antennae and swaying as they go. A worker moving along the tunnel would naturally tend to stray out of it from side to side as she goes. When the sense organs on her antennae tell her that the antenna has moved out of the tunnel, the worker will swing back. Amputation of one antenna leads to overcompensation on the other side. If the antennae are crossed over and fixed in that position, the worker finds great difficulty in orientating, but according to Hangartner she can still do so by a light compass response. All this gives no indication that directional information can be picked up from a trail, so can ants follow a trail in the right direction? It has been suggested that ants lay a trail which is polarised, for example in *M. ruginodis* (MacGregor, 1948). In this case, and also probably in *L. fuliginosus*, droplets of trail substance appear to be tapered so as to point one way. However, it is not clear how an ant following a 'vapour tunnel' trail, as described by Hangartner, could pick up this information. Some more recent experiments have indicated that directional clues come, not from the trail itself, but from other sources. For example, these clues come from other workers in the case of *Pheidologeton diversus* (Moffett, 1987). There is clearly scope for further contributions in this fascinating area. Hölldobler & Wilson (1990) list a wide range of chemicals used as trail substances. The techniques for isolating and identifying these substances require sophisticated and expensive equipment and are probably beyond the scope of readers of this book.

An excellent summary of this vast subject can be found in chapter 7 of Hölldobler & Wilson (1990).

Fig. 17. *Lasius fuliginosus* following a trail (see text) intact (a), or with one antenna missing (b), or with antennae crossed (c). (After Hangartner, 1967.)

Ants and plants

Ants and plants are associated at many levels, in all environments. The most intimate of these relationships are found in the tropics, where plants produce specialised feeding structures and sites for ants to build their nests in.

In Britain, relationships do occur, but they are much less intimate. In nearly all cases the association has some mutualistic elements.

Hölldobler & Wilson (1990) note the following strands:
ants protecting plants
plants sheltering ants
plants feeding ants
ants feeding plants
ants dispersing plants
ants pollinating flowers
ants pruning and weeding
ants parasitised by plants.

In nearly all situations ants will forage on plants and collect as prey insects, many of which are herbivorous (p. 24). In the wood ants there is no doubt that this removal of potential herbivores has a protective function. In Britain there was a reduction of 7% in the defoliation of trees in the presence of *Formica rufa* when compared with ant-free control trees (Skinner & Whittaker, 1981). This protection of trees by wood ants has been used as a form of biological control of economically important pest species. The impact of such ant predation on less numerous species of insect herbivore will, of course, be less obvious, but still worthy of investigation.

Plants as shelters

Worldwide there are many examples of ants living in foliage or other parts of plants. Some plants form special structures, called domatia. In Britain there are no such examples, but one ant species, *Lasius brunneus*, is known to nest in trees. This species is rarely seen on the ground and forages among the branches.

Some species of *Leptothorax*, especially *L. acervorum*, but also *L. nylanderi*, nest under bark, but usually in dead wood.

Plants as food

In the tropics there is a wide range of plants which produce special food bodies on which ants feed, but again no such associations seem to occur in Britain. However, plants produce food for ants in another way, in the extrafloral nectaries, and these are found in British plant species. The best known are those on bracken, *Pteridium aquilinum*.

Ants feeding plants

It is well known that more luxuriant plant communities occur close to ants' nests than in the surrounding areas. In Britain, King (1977) has made a careful study of vegetation near nests of *Lasius flavus*. The favourability of nests as sites for plant growth is probably due in part to the aeration of the soil and enrichment by nutrients from excretion and refuse dumping. However, this is not the whole story. King showed that some plant species are less frequent on the ant hills than elsewhere. Wood ants, too, have some effect on vegetation near their nests and also on soil pH (see p. 37) (Skinner, unpublished). Further study of the effects of nests on vegetation would be worthwhile.

Ants dispersing plants

Although British plants show disappointingly few adaptations to housing or feeding ants, many have seeds adapted for dispersal by ants. Some plant species have ant food bodies on the seed. These bodies attract ants which carry the seeds about, and thus disperse them. In Britain seeds of violets, *Viola,* and gorse, *Ulex,* are commonly carried about by *Lasius* and *Formica. Myrmica* is known to collect seeds of tormentil, *Potentilla erecta,* and store them in its nest, although we do not know what it does with them. In *Tetramorium* colonies, many seeds are fed to the brood. Beattie (1985) concluded that the main advantage to the plants of this seed transport comes from nutrient enrichment (the seeds are carried to nutrient-rich areas near the nest), and seed predator avoidance, but Hölldobler & Wilson (1990) point out that more data are needed to establish this conclusion with certainty. Observations of ants carrying seeds, and further research on the possible advantages to the seed and seedling, would be very valuable.

Ants as pollinators

There is only scant evidence that ants can act as important pollinators. It has been suggested that plants with a prostrate growth form and small flowers, together with a number of other features, would seem to be adapted to ant pollination (Hickman, 1974), but this idea is now in doubt. There is evidence that these plants are adapted for pollination by very small flying insects rather than by ants.

Recent studies have, indeed, shown that ants seem to impair pollen grain function; some pollen grains fail to germinate and those that do grow produce poor pollen tubes. Ants do not have specialised brood chambers, as bees and wasps do, and the suggestion has been put forward that they are vulnerable to attack by fungi and bacteria in the nest. A way of avoiding such attack would be to smear all brood with antibiotics. There is good evidence that ants produce these substances and that they have a negative effect on pollen grains. Thus, ants have evolved to produce antibiotics which protect them against microbial attack and, incidentally, affect pollen. This would exert a strong evolutionary pressure against flowers evolving adaptations for pollination by ants (Beattie, 1985; Peakall and others, 1991).

Ants pruning and weeding

In the tropics, some ants prune other plants overgrowing those on which they are nesting. Janzen (1967) has shown that his acacia ants did this. Real pruning or weeding have not been demonstrated in British ants, although local effects on vegetation have been noted near the nests of wood ants (Skinner, unpublished). The possibility that pruning or weeding is involved is worthy of exploration.

Ant guests, predators and parasites

One of the most fascinating aspects of the biology of ants is their association with other animals. There is no other comparable situation in the animal world, except, perhaps, situations involving human beings. Other social insects, the bees, wasps and termites, do have guests, but not in such variety as the ants. Virtually all of these ant guests are arthropods, although members of some other groups are found with ants occasionally.

The phenomenon of 'liking ants' (myrmecophily) is very complex, and embraces several types. An early classification of these, by Wasmann in the last century, is still used today, although it has some imperfections. Wasmann erected five categories of myrmecophile (Hölldobler & Wilson, 1990).

1. Synechthrans. The ants treat these in a hostile manner. They are predators of the ants and survive by effectively defending themselves, for example by running, or by defensive secretions.

2. Synoeketes. These are mainly scavengers which the ants ignore; presumably their neutral odour does not elicit a response from the ants.

3. Symphiles. These, sometimes called 'true guests', are more or less accepted by the ants as if they were part of the colony.

4. Ecto- and endoparasites and parasitoids. These are conventional parasites which live on or in their hosts.

5. Trophobionts. These are the insects that provide the ants with honeydew or glandular secretions and are in turn protected and nurtured by the ants. The honeydew producers have been dealt with in more detail in the section on feeding (p. 20).

The first four of these types belong to a larger grouping, referred to by Donisthorpe (1927) as intranidal guests (meaning guests within the nest). In all cases the ants are passive and are sought out by the guests, and the guests tend to be parasitic or commensal on the host ant, which generally suffers at the hands of the guest. The trophobionts are extranidal (living outside the nest). They are sought out by the ants, and the associations are of mutual benefit to both partners. The best known examples of this group are the honeydew-producing insects, which the ants milk (p. 20). In other cases, the myrmecophile produces secretions which the ants appear to crave. The best known of these are the caterpillars of blue butterflies.

British ants have a wide range of guests, and this short book cannot deal with them all in detail. Instead, we introduce a few common species of each kind to illustrate the principles and some of the gaps in our knowledge.

Intranidal guests

1. Synechthrans. In Britain, probably all ant guests of this type belong to the beetle family Staphylinidae (rove beetles), recognisable because their wing cases are much shorter than the abdomen (pl. 4). Rove beetles often hide in crevices in and near the nest, or beside trails. They distract attacking ants by appeasement behaviour in which the beetle thrusts its abdominal tip toward the attacking ant, and the ant licks it, usually losing interest in its attack. If this fails, the beetle can secret a further droplet from the abdomen tip which the ant then licks. When the ants are less aggressive the beetle may escape by feigning death. When it is in this state the ants may pick it up but they usually discard it later, unharmed. If all else fails, the beetle responds by curling the abdomen forward and thrusting it into the ant's face, at the same time releasing a repellent secretion. In addition, some of these beetles have a very ant-like appearance, which may help to protect them from the ants or from other enemies (Donisthorpe, 1927). One of the most abundant British species of this kind is *Myrmedonia humeralis*. Donisthorpe (1927) describes it as attacking and killing *F. rufa*, but Hölldobler & Wilson (1990) report research carried out in Japan which suggests that these beetles are largely scavengers. That work would bear repetition here. This beetle is also found with *Lasius fuliginosus*.

2. Synoeketes. A more advanced condition than that of *Myrmedonia* is found in *Dinarda*, another staphylinid beetle recorded with *Formica* in Britain. The larvae of this beetle feed on dead ants and debris and appease the ants in the way described for *Myrmedonia*. The adults roam around within the nest stealing prey brought back by the ants, and also, perhaps, feeding on mites, and the ants' eggs and larvae. They are known to steal food being passed from one worker to another and even to beg food from individual workers.

3. Symphiles. These are the so-called 'true' guests. According to Donisthorpe (1927), only five, staphylinid, beetle species fall into this category in Britain (table 5).

The clearest feature that distinguishes a species as a true guest is that it is fed by the ants with which it lives. This is true of all those in table 5. The commonest and most widespread of these true beetle guests in Britain is probably *Claviger testaceus*, although they are all restricted in range and rare in the north.

Table 5. *Beetles that are true guests of British ants (from Donisthorpe, 1927)*

Guest	Recorded hosts
Claviger testaceus	*Lasius flavus, L. niger, L. alienus, Myrmica scabrinodis*
C. longicornis	*Lasius niger, L. mixtus, L. umbratus*
Atemeles emarginatus	*Formica fusca* in summer; various *Myrmica* species in winter
A. paradoxus	*F. fusca* in summer; various *Myrmica* species in winter
Lomechusa strumosa	*Formica sanguinea*
(Amphotis marginata)	*Lasius fuliginosus*

(*Donisthorpe was not sure whether this beetle is a true guest or not.)

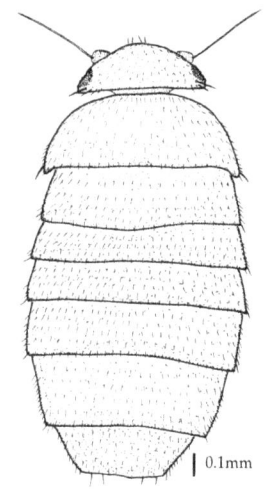

Fig. 18. A scuttle fly *A enigmatias* (Diptera, Phoridae) whose larvae are parasitic in ant pupae (from Disney, 1983).

4. Ecto- and endoparasites. Not much is known about the parasites of ants, although a number of fungal parasites have been identified around the world. None of these seems to be common, or of much importance in the colony, perhaps because of the ants' careful grooming and their secretion of fungicidal chemicals.

A number of animals living with ants can be regarded as ectoparasitic. Franks and others (1990) report on the obligate ectoparasitic mite *Antennophorus grandis*. This mite rides on the ant's body and either takes food that is being passed from worker to worker, or solicits it from its host worker. These authors showed that the mites greatly restrict the behaviour of the ants that bear them. Most of the observations were made on colonies kept in formicaria.

Donisthorpe mentions 39 species of mites associated with ants in Britain (table 6). Work on many of these others along the lines of the study by Franks and others would be worthwhile.

A whole group of insects, the endoparasitoids, specialise in laying eggs inside the bodies of others, which their larvae then consume from inside, killing the host in the process (fig. 18). Much remains to be found out about the endoparasitoids of British ant species. Table 7 summarises some of the knowledge we have so far.

Table 6. *Some mites associated with British ants (based on Donisthorpe (1927), in which more details may be found, and Vitzthum (1929, 1940–43))*

Species	Host(s)	Notes
Cilliba comata	*Lasius* species	2–3 specimens/ant
Urodiscella ricasoliana	*Lasius fuliginosus*	bright red
U. philoctena		attached to spur on fore-leg
Uroplitella minutissima	*Lasius niger, L. umbratus* and other species	many in galleries
Trachyuropoda coccinea	*Formica fusca, F. rufa*	
T. wasmanniana	*Lasius umbratus*	
T. bostocki	*Lasius umbratus, L. mixtus*	large species
T. canestriniana	*Tetramorium*	very rare
Uropolyaspis hamuliferus	*Lasius mixtus, L. umbratus, L. niger, L. flavus*	
Antennophorus grandis, A. uhlmanni, A. pubescens, A. foreli	*Lasius* species	
Gymnolaelaps myrmecophilus	*Myrmica scabrinodis*	
Laelaspis equitans	*Tetramorium*	
Holostaspis isotricha	*Formica rufa*	
Cosmolaelaps cuneifer	many species	
Sphaerolaelaps holothyroides	*Lasius umbratus, L. mixtus*	
Forcellinia wasmanni	*F. fusca, Myrmica ruginodis, M. rubra*	

Table 7. *Some of the known parasitoids of British ants*

Parasitoid	Ant host	Notes and references
A Hymenoptera		
Hybrizon buccata	'ants'	little is known (Gauld & Bolton, 1988)
Neoneurus species	'ants'	little is known (Shaw & Huddleston, 1991)
B Diptera (Phoridae, scuttle flies)		(see Disney, 1994)
Aenigmatias lubbocki	*Formica sanguinea, F. fusca, F. candida*	parasitic in ant pupae
Pseudacteon formicarum	*Lasius* species, *F. sanguinea Myrmica lobicornis*	parasitic in adult ants
P. brevicauda	*Myrmica* species	

Extranidal guests

Associations between ants and sap-sucking insects are described in more detail in the section on feeding (p. 19). Here, we consider the very interesting relationship between ants and the larvae of blue butterflies (family Lycaenidae) (table 8).

Table 8. *Lycaenid butterflies that have been found in the nests of British ants (more details are in Thomas & Lewington, 1991). More examples may be revealed by careful study of other lycaenid species*

Green hairstreak	*Callophrys rubi*
Purple hairstreak	*Quercusia quercus*
Silver-studded blue	*Plebejus argus*
Common blue	*Polyommatus icarus*
Chalkhill blue	*Lysandra coridon*
Adonis blue	*Lysandra bellargus*
Large blue	*Maculinea arion*

It is the large blue's association with ants that has been most widely studied in Britain, not least because the butterfly recently declined and went extinct in this country and a detailed study of its biology was needed as a basis for plans to reintroduce it. The butterfly larvae feed on the flowers of thyme, where they are attended by ants. When a larva reaches the fourth instar, it leaves its food plant and hides until it is found by a worker of the host ant. The ant then milks the caterpillar, which changes shape, retracting the head and expanding the thoracic segments to assume a humpbacked appearance (pl. 4). Perhaps mistaking it for an ant larva, the ant picks it up and takes it back to the nest. Here, the butterfly caterpillar feeds voraciously on the ant larvae, then pupates in the nest and emerges as an adult butterfly the next year. The natural history of this association is beautifully described by Thomas & Lewington (1991).

Predators

Some of the myrmecophiles dealt with above, such as the caterpillars of some Lycaenidae, are clearly predators on the ants or their brood. This section deals with some others. Amongst the arthropods not so far mentioned are a group of spiders in the family Theridiidae which sit on branches above trails, dropping on workers and wrapping them in silk. Ants are no doubt eaten by many generalist insect predators too, and much remains to be found out about the quantitative aspect of such predation.

Workers are eaten by a wide range of bird species, the best known being the green woodpecker *Picus viridis*. This bird can often be seen probing around in nests of many species during the winter. Its cigarette-butt-like droppings are also commonly seen on nests, and if pulled apart are found to be packed with indigestible ant exoskeletons. The number of ants eaten can be estimated by counting heads in the droppings. It is well known that many game birds eat ants and it was common practice in the past to introduce wood ants into woodland as food for pheasants. The wryneck *Jynx torquilla* also feeds on ants, but it is too rare to be an important predator in Britain.

Ants and slave ants: social parasitism

A number of species of ants depend on other species to a greater or lesser degree. The importance of temporary social parasitism in nest-founding has been discussed earlier (p. 17). The range of other possible relationships is illustrated in table 9.

Table 9. *Species relationships and social parasitism in ants (after Hölldobler & Wilson, 1990)*

Type	Details	British examples
A. Compound nests	species nest together, but keep brood separate	
plesiobiosis	species nest close together, but with little communication	many ants could be regarded as loosely associated
cleptobiosis	robber ants: small ants nest beside larger species and feed on refuse or rob returning workers	probably none
lestobiosis	thief ants: small ants share nests with larger species and steal food or prey on brood	*Solenopsis fugax*
parabiosis	two species nest together, and may share trails, but keep brood separate	none
xenobiosis	so-called guest ants inhabit nests of another species and solicit food but keep brood separate. These are probably parasites (although this is controversial)	*Formicoxenus nitidulus*
B. Mixed colonies	brood mingle and are cared for communally	
temporary social parasitism	a fertilised queen of one species enters the colony of another and assassinates the host queen, founding her own colony	see p. 17
dulosis (slavery)	one species captures brood of another to rear as slaves	*Formica sanguinea, Strongylognathus testaceus*
inquilinism (permanent parasitism)	parasite spends entire life in host nest; often workerless	*Anergates, Myrmica hirsuta, M. karavajevi*

Guest ants

Probably the only British guest ant is *Formicoxenus nitidulus*, which is found with *Formica rufa*. The colonies are small and the castes are not well differentiated. The ants remain in the nest of the host nearly all the time. They appear to feed by soliciting from the host or by stealing food from two mutually feeding host workers. The wood ants ignore the guests for most of the time, but occasionally attack. The response of the guest is to remain motionless, although in rare cases guests have been observed to sting the host. The relationship is almost a parasitic one, and indeed Hölldobler & Wilson (1990) consider it to be so, but Dumpert (1978) does not agree.

Thief ants

The only British thief ant is *Solenopsis fugax*. Its workers excavate narrow holes in the nest of the host. They then seize host brood, take them into the small holes where they cannot be followed, and eat them. They live with a wide range of species of *Formica* and *Lasius* in Britain.

Slavery

There are two slavemakers in Britain, the large wood-ant-like *Formica sanguinea* and *Strongylognathus testaceus*. *F. sanguinea* is thought to be a facultative slavemaker, as colonies are commonly found without slaves. Wheeler (1910) described a *F. sanguinea* raid as follows:

'The sorties occur in July and August after the marriage flight of the slave species has been celebrated and when only workers and mother queens are left in the formicaries. According to Forel the expeditions are infrequent, "scarcely more than two or three a year to a colony". The army of workers usually starts out in the morning and returns in the afternoon, but this depends on the distance of the *sanguinea* nest from the nest to be plundered. Sometimes the slavemakers postpone their sorties till three or four o'clock in the afternoon. On rare occasions they may pillage two different colonies before going home. The *sanguinea* army leaves its nest in a straggling, open phalanx sometimes a few metres broad and often in several companies or detachments. These move to the nest to be pillaged over the directest route over the often numerous obstacles in their path. As the forefront of the army is not headed by one or a few workers that might serve as guides, it is not easy to understand how the whole body is able to go so directly to the nest of the slave species, especially when this nest is situated, as is often the case, at a distance of 50 or 100 m...

When the first workers arrive at the nest to be pillaged, they do not enter it at once, but surround it and wait till the other detachments arrive. In the mean time the *fusca* or *rufibarbis* scent their approaching foes and either prepare to defend their nest or seize their young and try to break through the cordon of *sanguinea* and escape. They scramble up the grass-blades with their larvae and pupae in their jaws or make off over the ground. The sanguinary ants, however, intercept them, snatch away their charges and begin to pour into the entrances of the nest. Soon they issue forth one by one with the remaining larvae and pupae and start for home. They turn and kill workers of the slave-species only when these offer hostile resistance. The troop of cocoon-laden *sanguinea* straggle back to their nest, while the bereft ants slowly enter their pillaged formicary and take up the nurture of the few remaining young or await the appearance of future broods'.

Although the details of the raid are well known., colony founding in *F. sanguinea* has never been observed in nature. Laboratory observations, though, suggest that this is achieved by a form of temporary social parasitism, similar to that already described for *F. rufa* (p. 17).

Hölldobler and Wilson (1990) consider *Strongylognathus* to be evolving away from the slavemaking habit. The queens of host and parasite live side by side and raids have not been observed. The host queen supplies workers, and large colonies are generated. The parasite produces queens and workers. Perhaps *S. testaceus* is becoming an inquiline. If colonies can be found, some very interesting and original observations could be made. Bolton & Collingwood (1975) believe that *Strongylognathus* has a wider range than is indicated by the existing records, in the New Forest, Dorset and Devon.

Inquilinism

The inquiline ant species found in Britain are listed in table 9. Among a wide range of adaptations for inquilinism listed by Hölldobler & Wilson (1990), the most obvious is the loss of workers. All three of our species are workerless.

Anergates lives with *Tetramorium caespitum*. The strange males never develop fully and are described as pupoid for this reason. As far as we know, they fertilise the females in the natal nest. These females then fly off to find a new host colony and walk into it, usually unopposed. There they lay many eggs. The host appears to cease brood production after this and the *Tetramorium* colony eventually dies out.

Myrmica karavajevi is workerless too. It allows its host to continue to produce workers, but no sexual forms. The sexuals of *M. karavajevi* are winged and mate outside the nest.

Myrmica hirsuta has only been known for a few years and its biology is poorly understood.

Ants and the soil

Many ants live in the soil and build their nests there (p. 7). It seems likely, then, that they might have a significant effect on soil structure, just as earthworms do, particularly because of the sheer numbers of ants in certain habitats, and their habit of moving organic material about. The effects of ants on soils are both physical and chemical.

Physical effects

Studies in which active and inactive nest sites are compared with undisturbed soil nearby have concentrated on the effects of mound-building ant species. They show that ants transfer soil particles differentially, with the clays and silts being most commonly moved (Wiken and others, 1976). Baxter & Hole (1966) showed that ants (*Formica cinerea* in the USA) moved 7.4 metric tonnes per hectare of soil per year in mound-building operations. Is this of any great significance? Clearly, in the short term the movements are very localised near the nest, but many nests are short-lived and so in the longer term much of the habitat may be affected. Baxter & Hole estimated that *F. cinerea* could move a 15 cm layer of subsoil to the surface in a period of 106 years, on the reasonable assumption that they stayed in the same nest for about 10 years. In a related study, Salem & Hole (1968) credited ants with the conversion of a forest soil to a prairie soil.

In Europe, Czerwinski and others (1971) suggested that *Myrmica* species move to a fresh nest site twice, or occasionally more often, in a single season and thus spread their effects on soil further than one might at first suppose. A note of caution comes from another American study in which Wiken and others (1976) showed that soil in unoccupied mounds was very similar to that in undisturbed areas, and so concluded that the ant effect is only short-lived.

Chemical effects

Do ants affect soil pH? Many species produce formic (methanoic) acid as a defensive or alarm substance (p. 12). In Britain, research has focused on the mound-building ant *Lasius flavus*. Most of the studies indicate no significant difference in pH between mounds of this species and the surrounding soil, except in acidic grassland where ant activity seems to increase pH. There is no evidence at all for the expected decrease in pH. The increase of pH in acidic grassland is probably caused by accumulation of basic material in nests, reducing leaching or allowing retention of basic cations. If ants affect soil pH, does this in turn affect vegetation? The answer seems to be no.

It has been suggested that ants affect the nutrient status of the soil, and here there is more evidence. The most frequently reported effects involve phosphorus, potassium and carbon. In *Myrmica* nests these elements have been shown to be increased by as much as four times. Simple chemical soil analysis kits are easily obtainable and much useful work could be done with these in and around nests of British species.

3 Identification

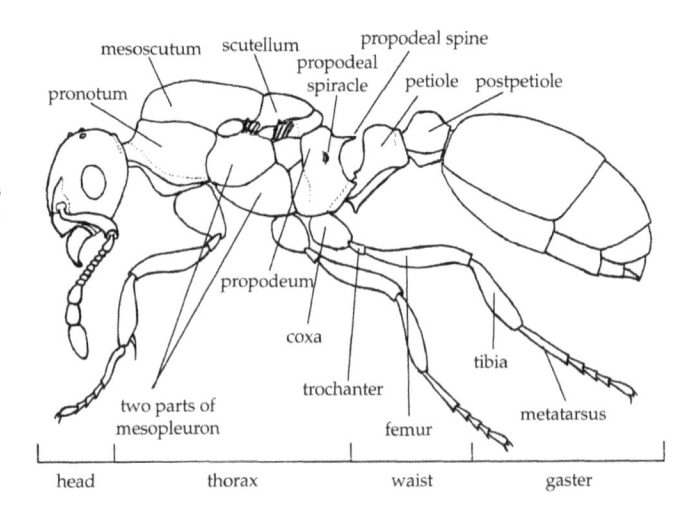

Glossary diagram 1. Side view of *Leptothorax tuberum* queen that has shed her wings. The thorax of male ants is usually similar in structure to that of queens. The abdomen of the queen is large and accommodates the enlarged ovaries.

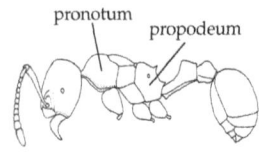

Glossary diagram 2. Worker of *Stenamma debile*. The plates on the thorax are reduced, as there are no flight muscles.

Introduction

Keys I–III are designed to enable beginners to identify British species of ants. Exotic species that have established in hothouses and heated buildings are also described at the relevant points in the worker key. These simple keys will make it possible to name most species found in Britain, but not all of them. Some groups of species have always been tricky to determine and often the novice is required to refer the more difficult specimens to a specialist. Examples are in the yellow species of *Lasius*, the wood ants and the *Myrmica scabrinodis* group. Indeed even experts cannot reliably identify some species in the worker caste unless a whole series of specimens is available from the same nest. As with any taxonomic work, the treatment of the classification of the British ants given here is a snap-shot in time. There have been several recent revisions, adding species and changing names, and there are likely to be more in the near future.

A few unfamiliar terms are needed to describe body parts. These are defined in glossary diagrams 1–3. The part of the body to which the legs are attached is called the thorax here but it is more correctly known as the 'alitrunk'. Technically, it includes the first segment of the true insect abdomen, a segment termed the propodeum in the higher Hymenoptera, to which the ants belong. We have called the following one or two narrowed segments the waist, which many other authors call the 'pedicel'. The last body section is the gaster.

Glossary diagram 3.
Head of *Lasius niger* worker.

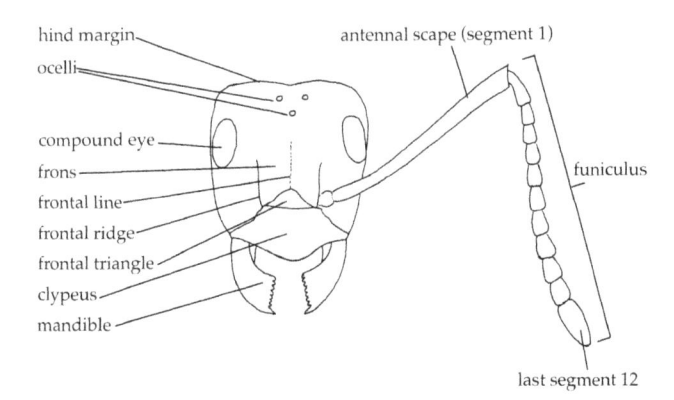

Labels: hind margin, ocelli, compound eye, frons, frontal line, frontal ridge, frontal triangle, clypeus, mandible, antennal scape (segment 1), funiculus, last segment 12

Keys I–III have been designed for use with a low power dissecting microscope, with a magnification of at least 20x, preferably 40x or more. A hand-held lens may not give quite enough magnification. Some keys to ants require palpal segments of the mouthparts to be measured and sting characters to be examined, for which the specimen has to be carefully mounted and dissected to display the various features. Our keys do not require such dissection but it is advisable to open the mandibles of the specimens while they are still relaxed, so that the number and shape of the teeth on the mandibles can be seen.

There are three separate keys: to workers, queens and males. Before you can key down a specimen you need to check that it is an ant, by referring to p. 1 or Yeo & Corbet (1995), and determine its sex and caste. Males can be reliably recognised by the complex genital capsule at the tip of the gaster. In most British species they are winged and short-lived. Females (queens and workers) have defensive organs at the tip of the gaster. The myrmicine and ponerine females possess stings, whilst dolichoderine and formicine females have no stings but release chemical secretions made in glands in the gaster. Dolichoderinae produce repellent substances from anal glands, whilst in Formicinae the poison gland produces a mixture dominated by formic acid. Workers are always wingless whereas the queens have wings at first but shed them after mating. The tiny vestiges of the wing bases remain visible on the thorax. In a small minority of species the male is wingless and may then closely resemble the worker of the same species.

The Quick-check field key (p. 58) deals with only the very commonest species, and may be used as a time-saving reminder for beginners learning the common species. It cannot be used for serious identification.

Keys are also provided by Bolton & Collingwood (1975) and Brian (1977). A simple key to the common species described on p. 60 is to be found in Skinner (1987).

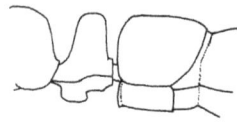

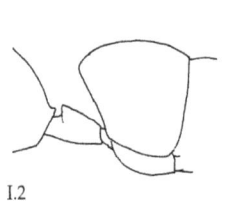

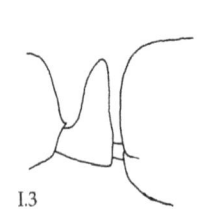

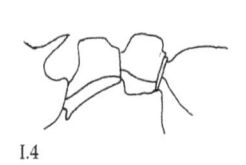

I Worker ants

I.1

1   Waist consisting of only one segment, the petiole (I.1, 2 & 3)   2
–   Waist consisting of two segments, the petiole and postpetiole (I.4)   Myrmicinae 24

I.2

2   Gaster with a distinct constriction between the first and second segments (I.1); sting present
                                    Ponerinae *Ponera coarctata*

Hypoponera punctatissima also keys out here. It is a hot-house species accidentally imported by commerce and can be distinguished by the structure of the underside of the petiole, which lacks a backward pointing projection found in *Ponera coarctata* (shown in I.1). *Hypoponera punctatissima* has a wingless, worker-like male. There are a few records of this species outside of heated buildings but these are exceptional.

–   Gaster with no constriction between the first and second segments (I.2); sting absent   3

I.3

3   Petiole very small, overhung by the first segment of the gaster (I.2), not visible from above
                              Dolichoderinae, genus *Tapinoma* 4
–   Petiole well developed, scale-like (I.3), clearly visible from above   Formicinae 5

The hothouse tramp species *Linepithema humile*, a dolichoderine, keys out here but is distinguished by the tip of the gaster, which has a groove across it typical of Dolichoderinae, rather than a conical pore surrounded by a tuft of hairs as in the Formicinae.

I.4

4   Notch in clypeus as shallow as I.5; ant generally smaller   *Tapinoma ambiguum* (pl. 6.3)
–   Notch in clypeus as deep as I.6; ant generally larger
                                           Tapinoma erraticum

These two species of *Tapinoma* are very similar and cannot always be separated unless the specimens are males. There is an introduced, hothouse species of *Tapinoma*, *T. melanocephalum*, smaller than either of the native species and with pale gaster and legs.

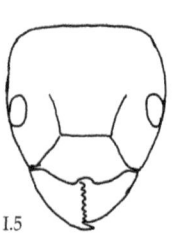

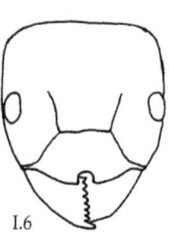

I.5   I.6

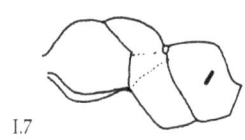

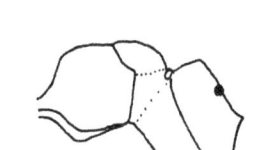

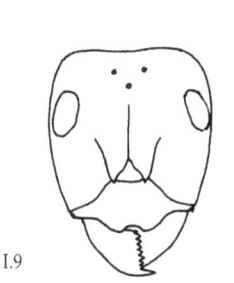

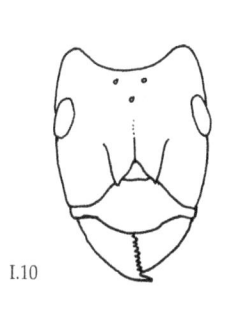

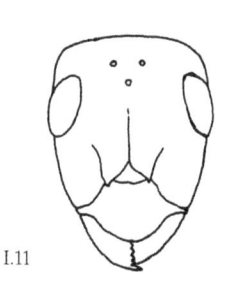

5   Opening of propodeal spiracle (breathing pore) elongate, slit-like, on side face of propodeum (I.7); hind tibia with a double row of bristles present on the underside (in addition to those at the tip) genus *Formica* 6
–   Opening of propodeal spiracle circular or nearly so, on curved surface between side and hind faces (I.8); hind tibia with no rows of bristles on the underside, although there are bristles at the tip (where there are also usually abundant, very short, soft, hairs pressed flat against the tibia) genus *Lasius* 16

Three formicine genera – *Camponotus* (imported), *Paratrechina* (imported) and *Plagiolepis* (one Channel Islands species) – are not treated here. *Camponotus* species tend to be large ants, with two or more forms of worker, often nesting in wood. Two species of *Paratrechina* (*P. vividula* and *P. longicornis*) are found in heated buildings such as hothouses. *Plagiolepis vindobonensis* is distinguished by having 11-segmented antennae in the worker and queen.

6   Front margin of clypeus with a notch in the middle (I.9) *Formica sanguinea* (pl. 2.1)
–   Front margin of clypeus without a notch (e.g. I.10, 11) 7

7   Head strongly concave behind (I.10) *Formica exsecta* (pl. 1.5)
–   Head not strongly concave behind (I.11) 8

8   Thorax dark, the same colour all over 9
–   Thorax with reddish areas, which may be confined to the junctions between the thoracic segments 11

9   Strongly shining; black or very occasionally brownish-black; lower surface of head with two long hairs (these are occasionally absent in old, abraded specimens) *Formica candida*
–   Not strongly shining; black, occasionally brownish-black; lower surface of head without hairs 10

10 Numerous erect hairs on top of pronotum; femora of middle legs with a few long hairs beneath *Formica lemani*
–   Few (1–3), if any, erect hairs on top of pronotum; femora of middle legs without long hairs *Formica fusca* (pl. 6.5)

11 Frontal triangle very finely roughened, dull, not reflecting light 12
–   Frontal triangle reflecting light or shining, unsculptured or with a few very fine pits or with a crescent-shaped area of quite dense, fine pits across it 13

The wood ants: a difficult group. Consult an expert if in doubt.

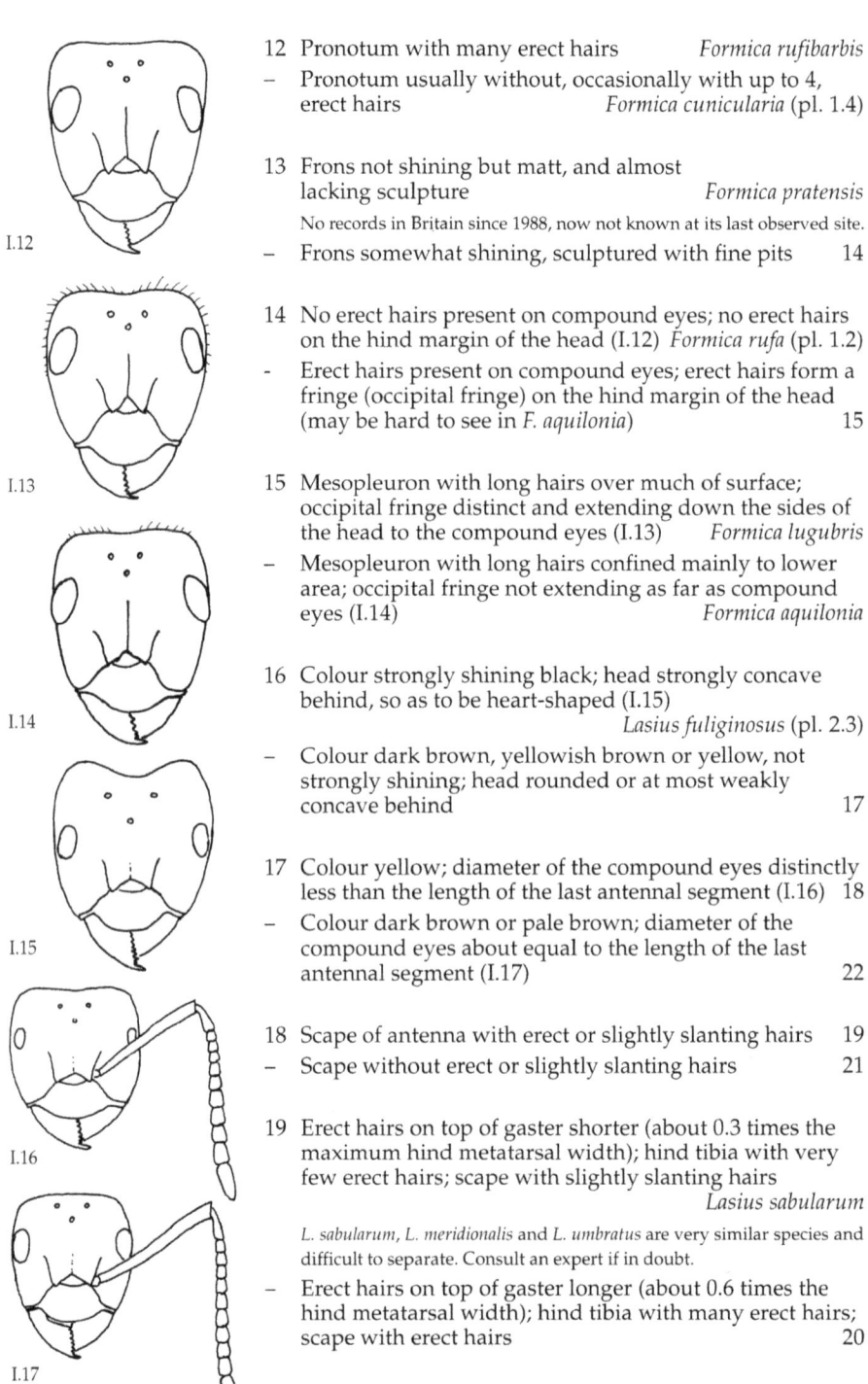

12 Pronotum with many erect hairs   *Formica rufibarbis*
 − Pronotum usually without, occasionally with up to 4,
    erect hairs   *Formica cunicularia* (pl. 1.4)

13 Frons not shining but matt, and almost
    lacking sculpture   *Formica pratensis*
    No records in Britain since 1988, now not known at its last observed site.
 − Frons somewhat shining, sculptured with fine pits   14

14 No erect hairs present on compound eyes; no erect hairs
    on the hind margin of the head (I.12) *Formica rufa* (pl. 1.2)
 − Erect hairs present on compound eyes; erect hairs form a
    fringe (occipital fringe) on the hind margin of the head
    (may be hard to see in *F. aquilonia*)   15

15 Mesopleuron with long hairs over much of surface;
    occipital fringe distinct and extending down the sides of
    the head to the compound eyes (I.13)   *Formica lugubris*
 − Mesopleuron with long hairs confined mainly to lower
    area; occipital fringe not extending as far as compound
    eyes (I.14)   *Formica aquilonia*

16 Colour strongly shining black; head strongly concave
    behind, so as to be heart-shaped (I.15)
                                               Lasius fuliginosus (pl. 2.3)
 − Colour dark brown, yellowish brown or yellow, not
    strongly shining; head rounded or at most weakly
    concave behind   17

17 Colour yellow; diameter of the compound eyes distinctly
    less than the length of the last antennal segment (I.16) 18
 − Colour dark brown or pale brown; diameter of the
    compound eyes about equal to the length of the last
    antennal segment (I.17)   22

18 Scape of antenna with erect or slightly slanting hairs   19
 − Scape without erect or slightly slanting hairs   21

19 Erect hairs on top of gaster shorter (about 0.3 times the
    maximum hind metatarsal width); hind tibia with very
    few erect hairs; scape with slightly slanting hairs
                                               Lasius sabularum
    L. sabularum, *L. meridionalis* and *L. umbratus* are very similar species and
    difficult to separate. Consult an expert if in doubt.
 − Erect hairs on top of gaster longer (about 0.6 times the
    hind metatarsal width); hind tibia with many erect hairs;
    scape with erect hairs   20

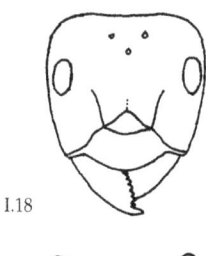

Key I Worker ants

I.18

I.19

I.20

I.21

I.22

20 Scape and tibiae somewhat flattened in cross section; at mid-point of scape maximum width about twice minimum width *Lasius meridionalis*
- Scape and tibiae almost circular in section; at mid-point of scape maximum width about one and a half times minimum width *Lasius umbratus* (pl. 2.4)

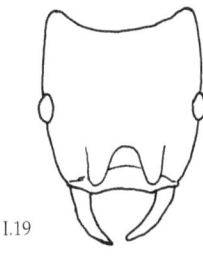

21 Hairs on upper surface of gaster longer (about 0.6 times the maximum hind metatarsal width); no erect hairs on underside of head *Lasius flavus* (pl. 9.3)
- Hairs on upper surface of gaster shorter (about 0.3 times the maximum hind metatarsal width); erect hairs present on underside of head *Lasius mixtus*

22 Frontal triangle well defined (I.18); scale of petiole notched; gaster darker than thorax and petiole
Lasius brunneus
- Frontal triangle poorly defined (I.17); scale of petiole not notched; gaster and thorax fairly uniform dark brown 23

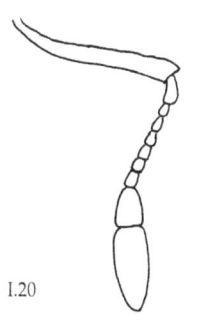

23 Scape of antenna and hind tibia with standing hairs
Lasius niger (pl. 9.1)

Seifert (1992) has shown that the species previously known as *Lasius niger* comprises two species in the UK, true *L. niger* and the very similar *L. platythorax*. The worker of *L. platythorax* is said to have more dense hairs on the antennae and legs and sparser body hair, best noted on the clypeus. Consult an expert if in doubt. The Channel Islands species *Lasius emarginatus* is similar to *L. niger* and *L. platythorax* but has slanting antennal hairs and the thorax and petiole are yellowish-red.

- Scape and hind tibia without erect hairs *Lasius alienus*

Seifert (1992) recognises *L. psammophilus* as a species distinct from *L. alienus*. Consult an expert if a species identification is important.

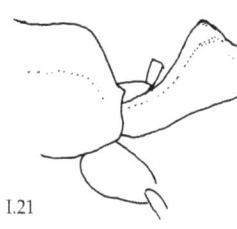

24 Mandibles long, slender, curved and toothless (I.19)
Strongylognathus testaceus (pl. 6.1)
- Mandibles shaped as in I.24 & 25 and with at least 5 teeth 25

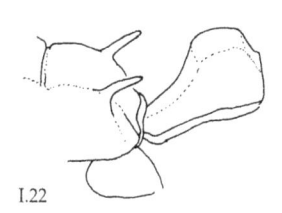

25 Antennae with 10 segments, the two at the tip forming a club (I.20); propodeum without a pair of teeth or spines (I.21) *Solenopsis fugax* (pl. 6.2)

A second species of *Solenopsis*, *S. monticola*, occurs in the Channel Islands, as does *S. fugax*. The introduced species *Monomorium pharaonis* and *M. salomonis* key out here on the basis of the spineless propodeum but have antennae with 12 segments, the three at the tip forming a club. They are found in heated buildings.

- Antennae with 11 or 12 segments; propodeum with a pair of teeth or spines (I.22) 26

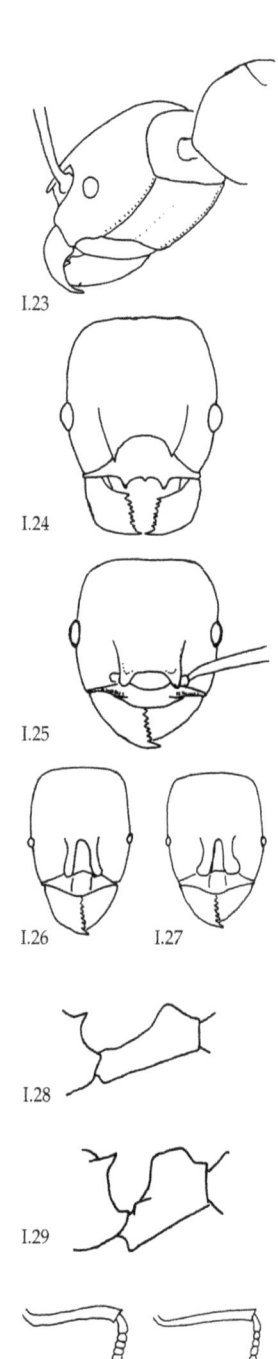

Crematogaster scutellaris keys out here but is an introduced species with 11-segmented antennae. It is very distinct; the petiole has no node and the postpetiole is attached to the upper surface of the gaster, which is flattened. The gaster is held curved above the thorax when the ant is alarmed. This species is often associated with imported cork from south Europe.

26 Lower surface of head with a pair of ridges along it, one each side (I.23); front margin of clypeus with two prominent teeth and sometimes with a third small tooth in the middle (I.24) *Myrmecina graminicola* (pl. 7.3)

— Lower surface of head without a pair of ridges; front margin of clypeus without teeth 27

27 Sides of the clypeus with their hind margins raised into a transverse ridge in front of the antennal bases (I.25)
    Tetramorium caespitum (pl. 5.4)

There are four hot-house species of *Tetramorium* recorded from Britain. They can be separated from the native *T. caespitum* because the frontal ridges extend back towards the crown of the head, beyond the level of the compound eyes.

— Sides of the clypeus not raised into a transverse ridge 28

28 Middle portion of clypeus with two small ridges along it (I.26 & 27); petiole elongate in front, as in I.28
    genus *Stenamma* 29

Two other species key out here on the basis of the petiole shape but do not have the ridges on the clypeus: *Pheidole megacephala* (imported, in heated buildings) and *Aphaenogaster subterranea* (Channel Islands). *P. megacephala* is a small ant with workers of two very different forms (a large-headed soldier and a more typical, small-headed worker) whilst *A. subterranea* is a moderately large, long-legged ant in a genus more typical of warmer climates.

— Middle portion of clypeus without a pair of ridges; petiole not elongate in front (I.29) 30

29 Width of central smooth area of frons (frontal triangle) greater than one third the total distance between bases of antennae (I.26) (ant with shorter, stouter legs)
    Stenamma debile (pl. 7.4)

— Width of central smooth area of frons (frontal triangle) less than one third the total distance between bases of antennae (I.27) (ant with longer, narrower legs)
    Stenamma westwoodii

These two species of *Stenamma* are very difficult to separate. Consult an expert if in doubt. Males are easier to identify than workers and queens.

30 The last three antennal segments as long as the remainder of the funiculus (I.30); mandible with 5 or 6 teeth 31

— The last three antennal segments distinctly shorter than the remainder of the funiculus (I.31); mandible with occasionally 6, usually 7 or more teeth genus *Myrmica* 35

PLATE 1

1. *Formica lugubris* ♂

2. *Formica rufa* ♀
   (from above)

3. *Formica rufa* ♀
   (from left side)

4. *Formica cunicularia* ♀
   (dark form)

5. *Formica exsecta* ♀

6. *Formica rufa* ♀

Scale lines 3 mm

♂ = male
♀ = queen
☿ = worker

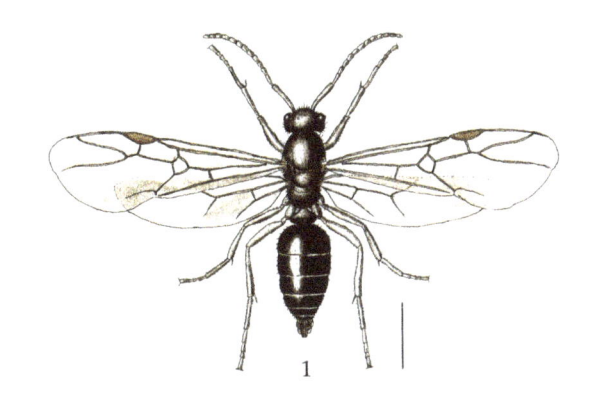

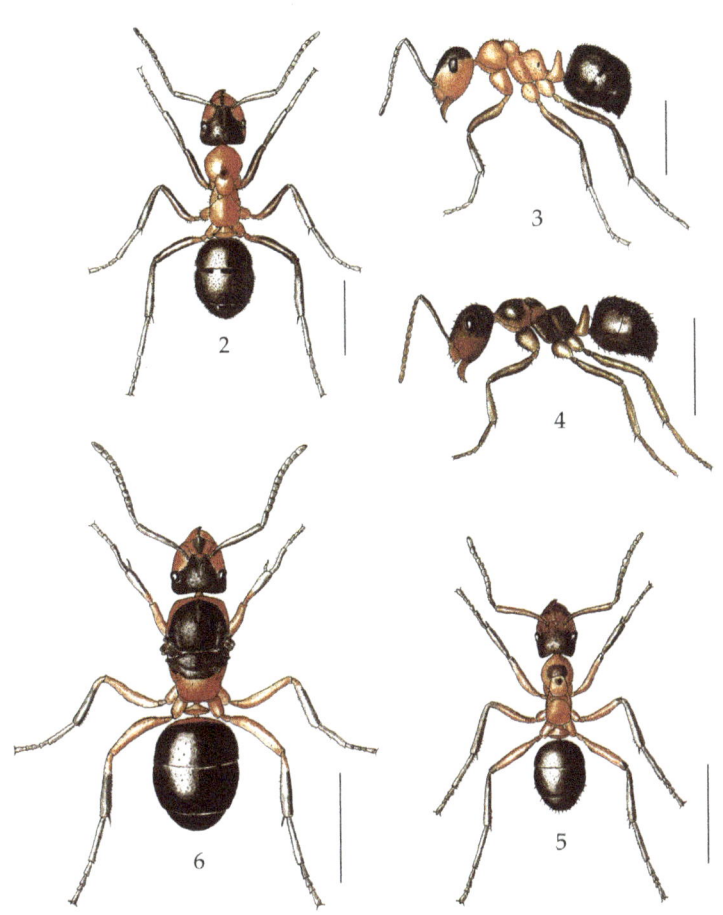

PLATE 2

1. *Formica sanguinea* ♀

2. *Formica sanguinea* monster (left half male, right half worker)

3. *Lasius fuliginosus* ♀

4. *Lasius umbratus* ♀

5. *Hypoponera punctatissima* ♂ (an introduced species)

6. *Tetramorium caespitum* ♂

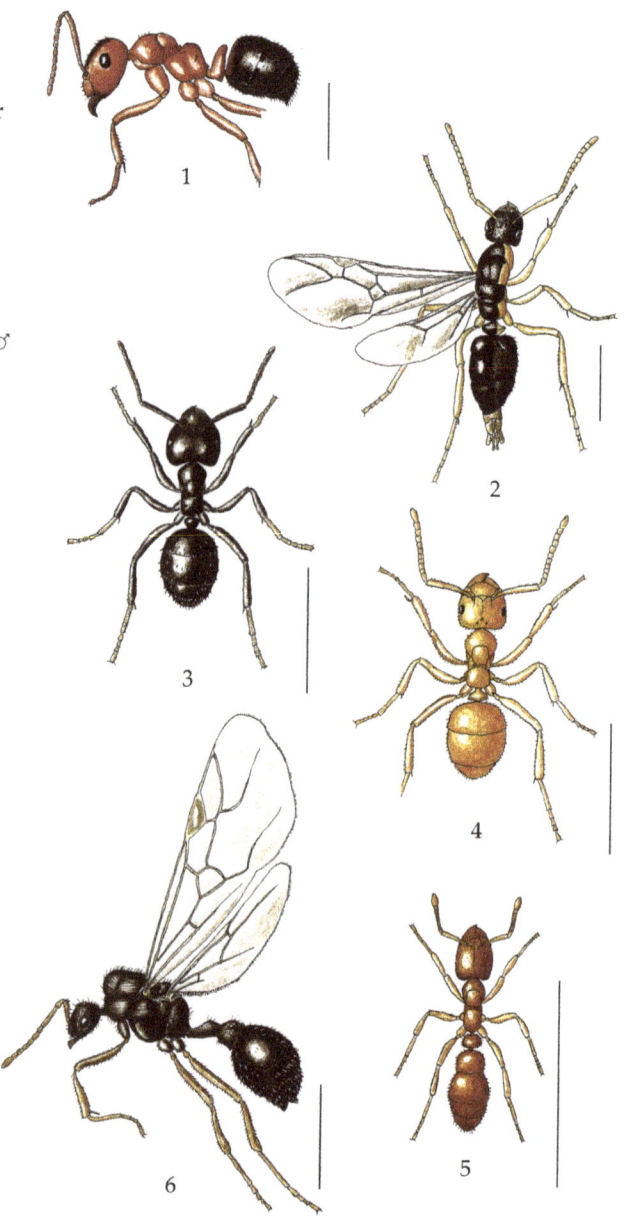

Scale lines 3 mm

PLATE 3

1. *Leptothorax interrruptus* ♀
2. *Myrmica ruginodis* ♀
3. *Myrmica sulcinodis* ♀
4. *Myrmica ruginodis* macro queen
5. *Myrmica hirsuta* ♀ (a social parasite)
6. *Myrmica ruginodis* micro queen

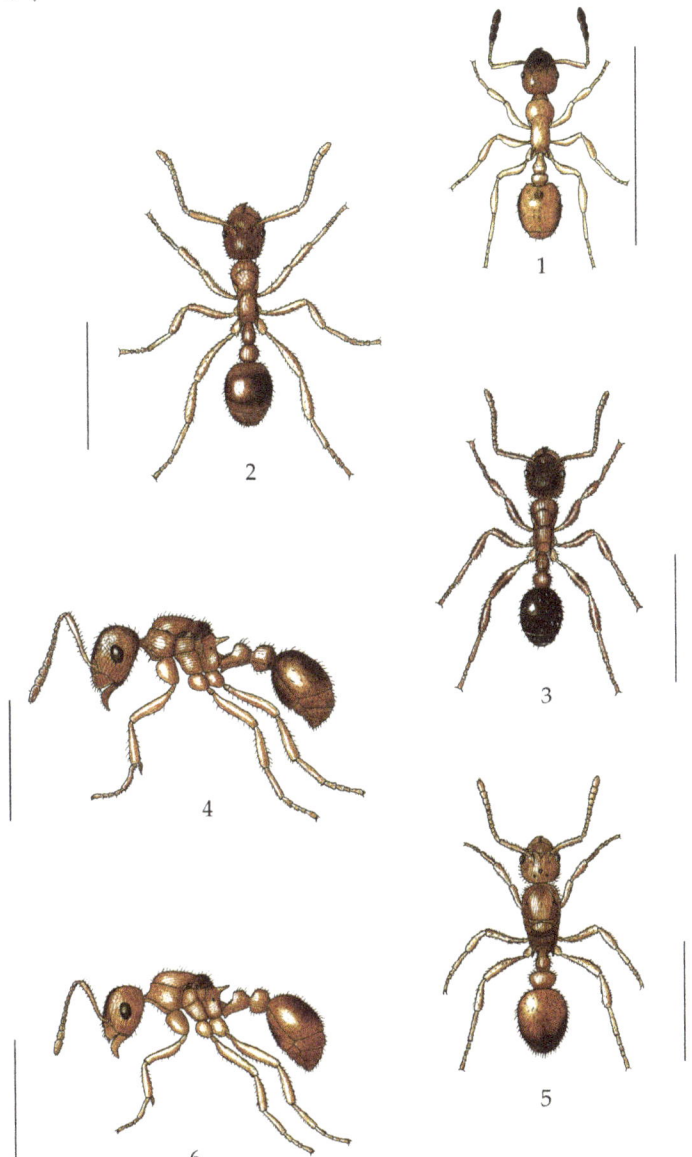

Scale lines 3 mm

PLATE 4

Ant brood and ant guests

1. *Lasius niger*
   eggs and first, second and fourth stage larvae

2. *Lasius niger*
   worker (left) and queen pupal cocoons

3. *Myrmica* species
   queen pupa

4. The beetle
   Claviger testaceus

5. The beetle
   Atemeles pubicollis

6. A caterpillar of the large blue butterfly
   Maculinea arion

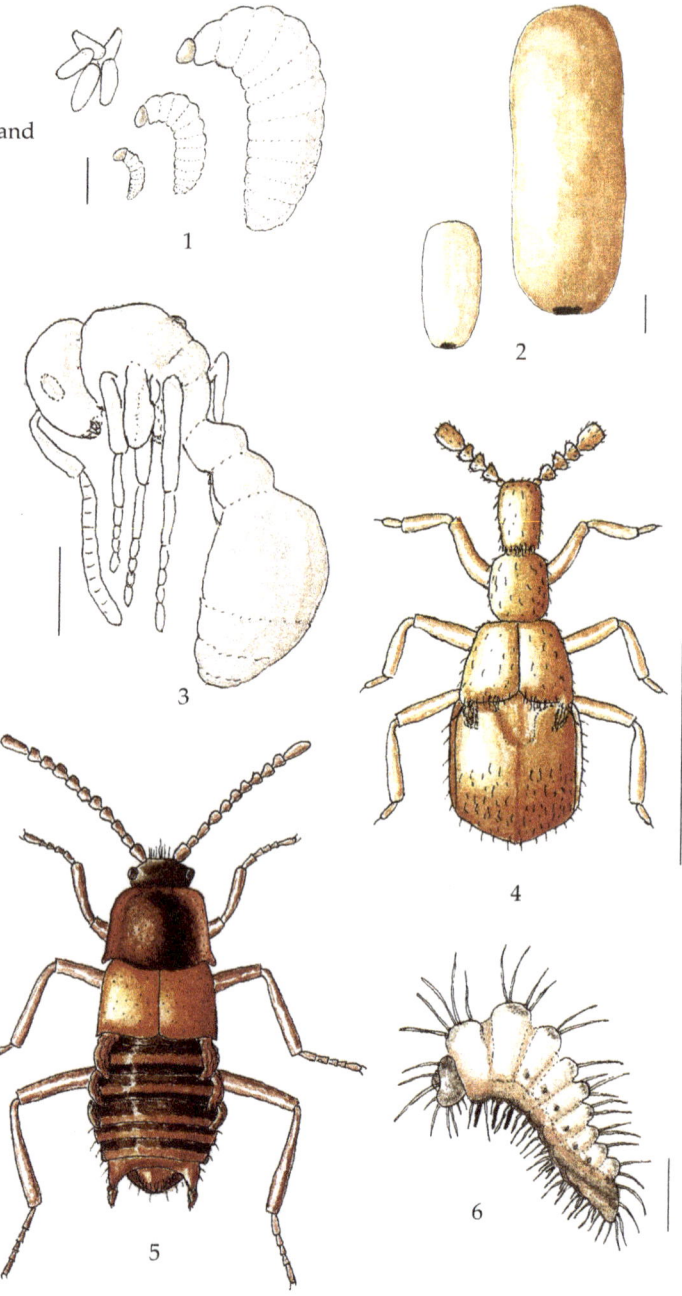

Scale lines 1 mm

PLATE 5

1. *Leptothorax tuberum* ♂
2. *Leptothorax acervorum* ♀
3. *Formicoxenus nitidulus* ♀
4. *Tetramorium caespitum* ♀

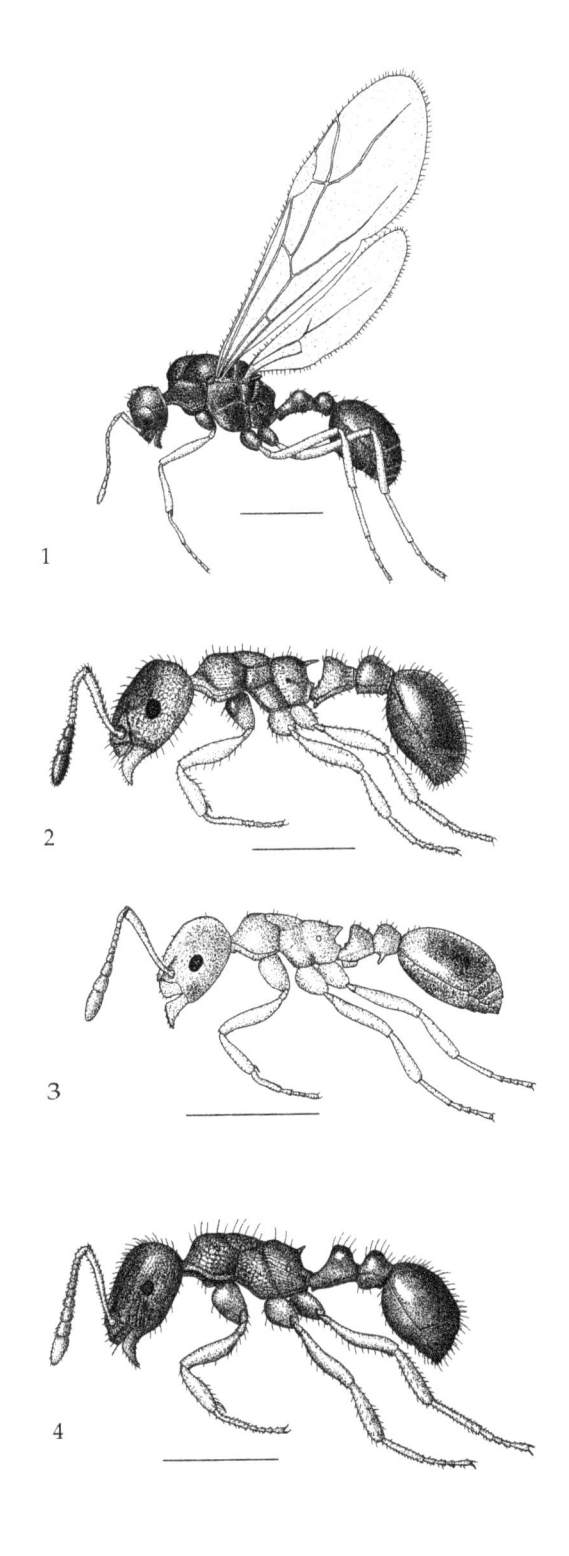

Scale lines 1 mm

PLATE 6

1. *Strongylognathus testaceus* ♀
2. *Solenopsis fugax* ♀
3. *Tapinoma ambiguum* ♀
4. *Formica aquilonia* wingless ♀
5. *Formica fusca* ♀

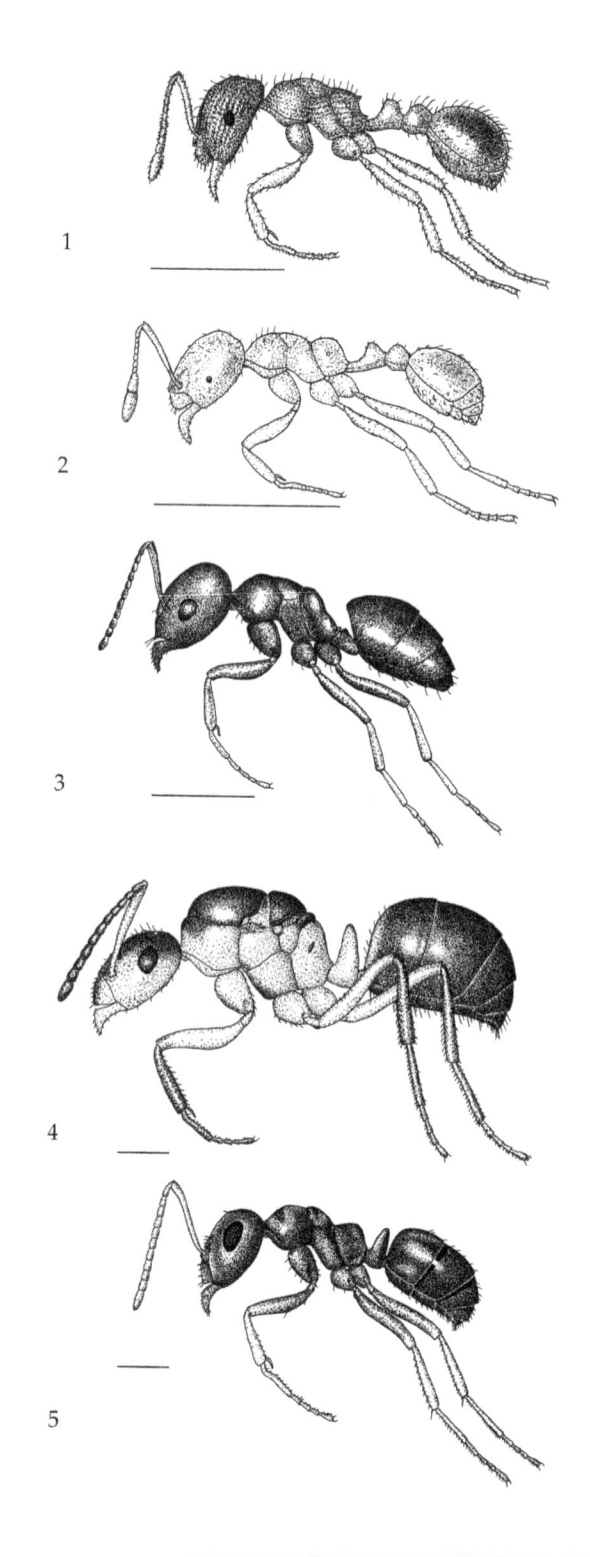

Scale lines 1 mm

PLATE 7

1. *Ponera coarctata* ⚥
2. *Ponera coarctata* ♂
3. *Myrmecina graminicola* ⚥
4. *Stenamma debile* ⚥

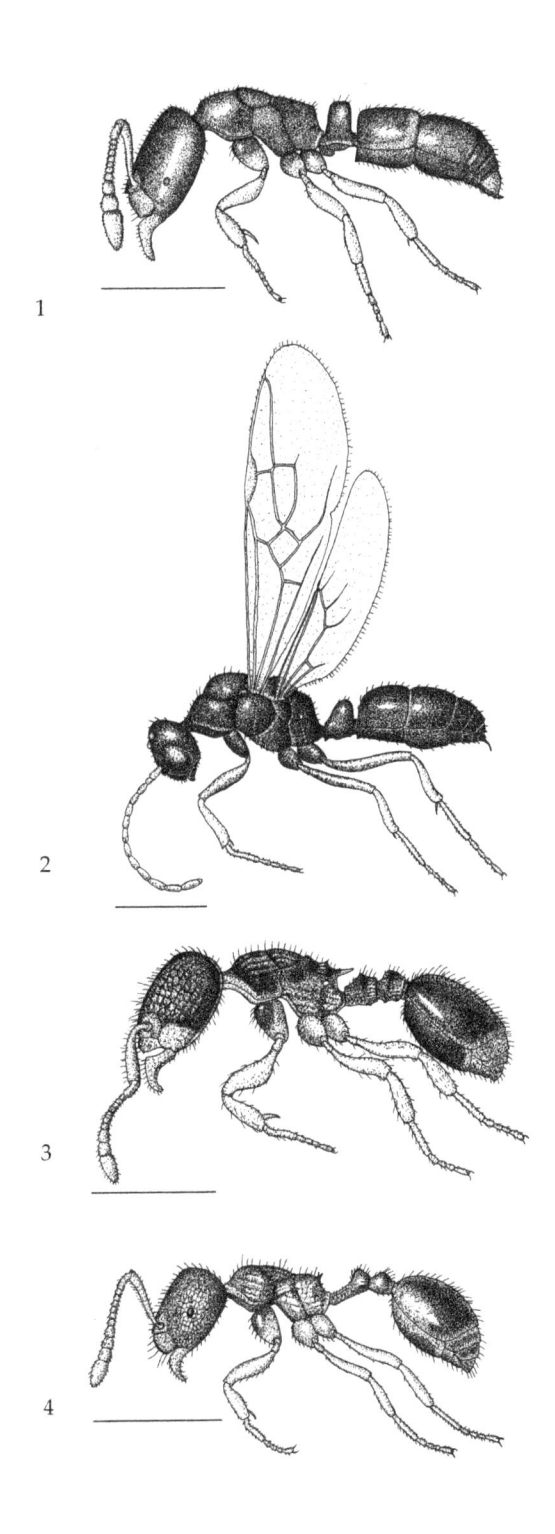

Scale lines 1 mm

PLATE 8

1. *Myrmica rubra* winged ♀
2. *Myrmica scabrinodis* ♀
3. *Anergates atratulus* wingless ♀

   dorsal view showing channel in gaster

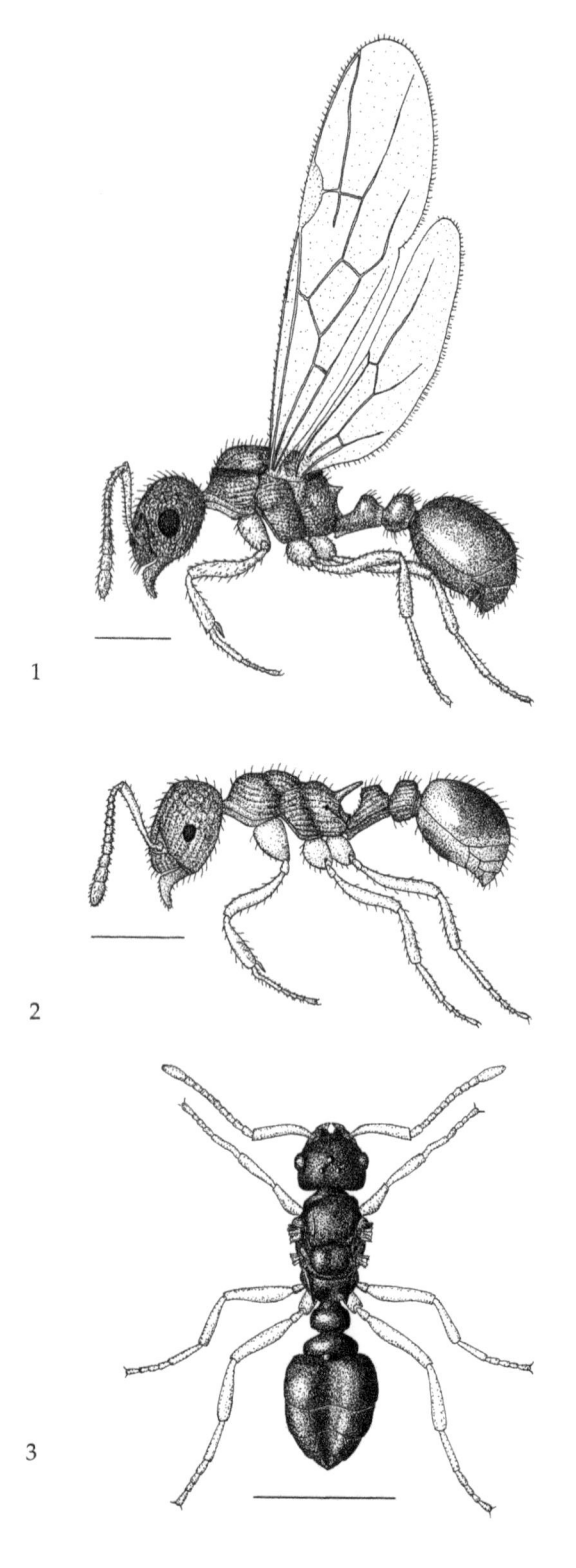

Scale lines 1 mm

PLATE 9

1. *Lasius niger* ♀
2. *Lasius flavus* ♂
3. *Lasius flavus* ♀
4. *Lasius fuliginosus* ♀

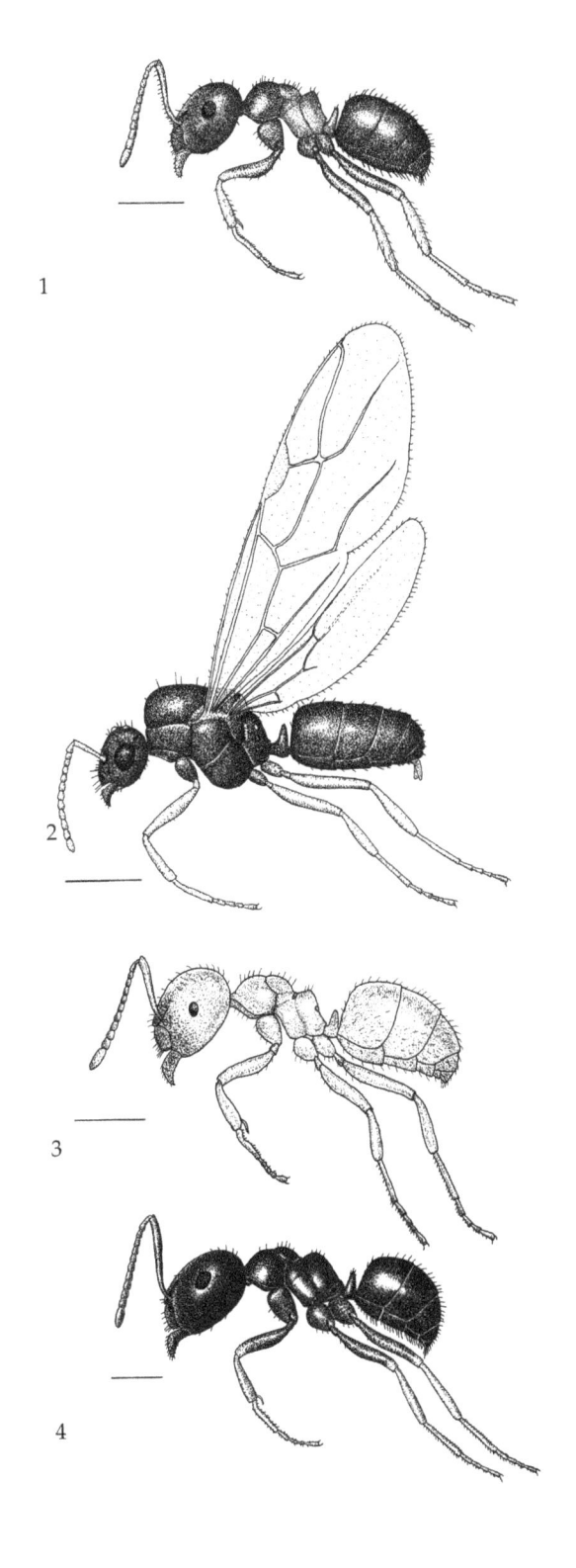

Scale lines 1 mm

PLATE 10

Wing maps

1. *Ponera coarctata* ♀
2. *Tetramorium caespitum* ♀
3. *Myrmecina graminicola* ♀
4. *Strongylognathus testaceus* ♀
5. *Myrmica rubra* ♀
6. *Anergates atratulus* ♀
7. *Leptothorax acervorum* ♂
8. *Tapinoma erraticum* ♂
9. *Formica exsecta* ♀
10. *Lasius fuliginosus* ♀

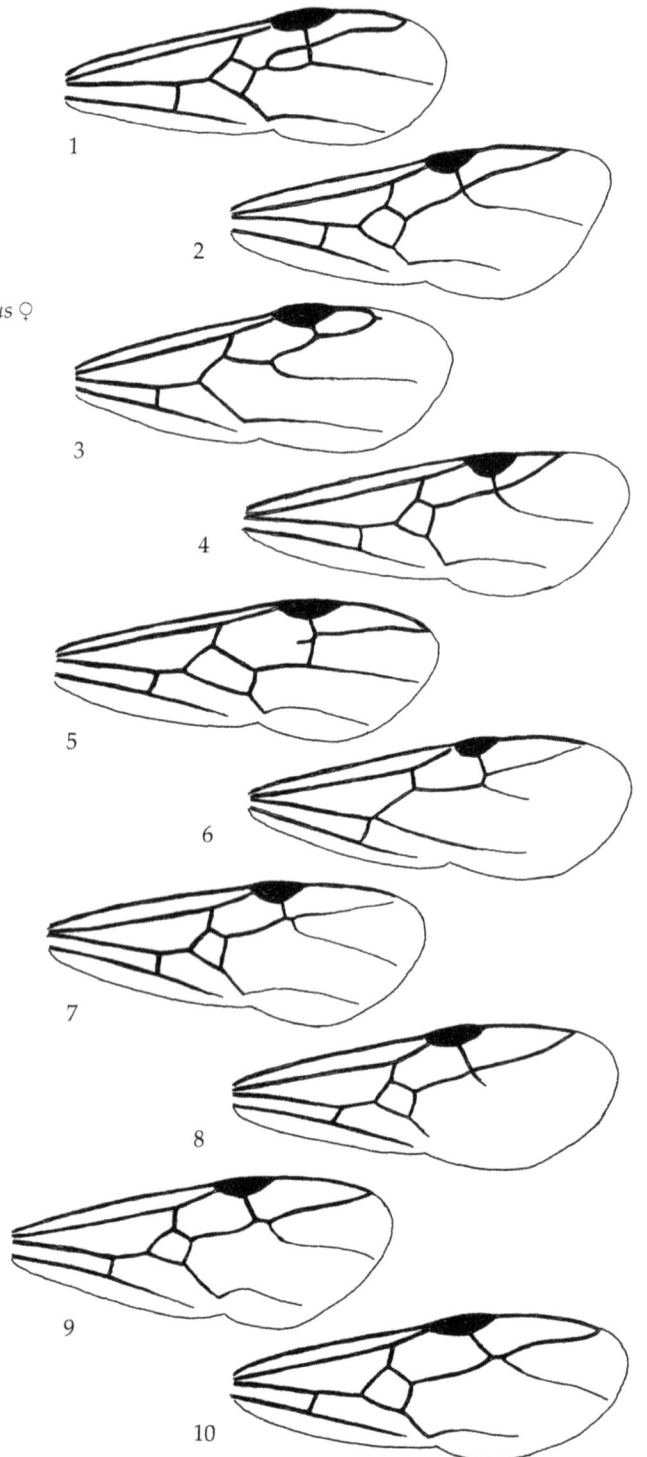

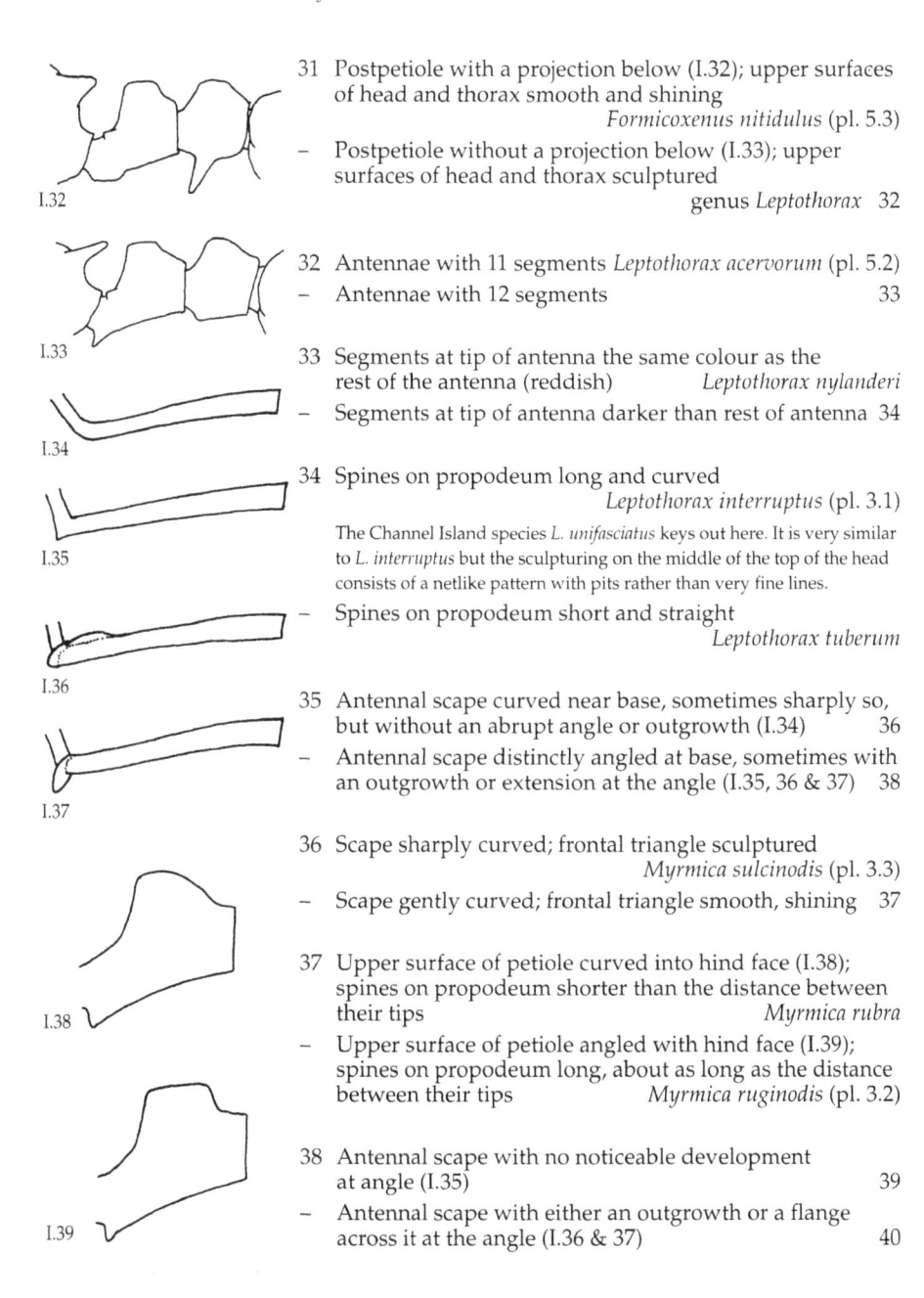

31 Postpetiole with a projection below (I.32); upper surfaces of head and thorax smooth and shining
Formicoxenus nitidulus (pl. 5.3)
- Postpetiole without a projection below (I.33); upper surfaces of head and thorax sculptured
genus *Leptothorax* 32

32 Antennae with 11 segments *Leptothorax acervorum* (pl. 5.2)
- Antennae with 12 segments 33

33 Segments at tip of antenna the same colour as the rest of the antenna (reddish) *Leptothorax nylanderi*
- Segments at tip of antenna darker than rest of antenna 34

34 Spines on propodeum long and curved
Leptothorax interruptus (pl. 3.1)

The Channel Island species *L. unifasciatus* keys out here. It is very similar to *L. interruptus* but the sculpturing on the middle of the top of the head consists of a netlike pattern with pits rather than very fine lines.

- Spines on propodeum short and straight
Leptothorax tuberum

35 Antennal scape curved near base, sometimes sharply so, but without an abrupt angle or outgrowth (I.34) 36
- Antennal scape distinctly angled at base, sometimes with an outgrowth or extension at the angle (I.35, 36 & 37) 38

36 Scape sharply curved; frontal triangle sculptured
Myrmica sulcinodis (pl. 3.3)
- Scape gently curved; frontal triangle smooth, shining 37

37 Upper surface of petiole curved into hind face (I.38); spines on propodeum shorter than the distance between their tips *Myrmica rubra*
- Upper surface of petiole angled with hind face (I.39); spines on propodeum long, about as long as the distance between their tips *Myrmica ruginodis* (pl. 3.2)

38 Antennal scape with no noticeable development at angle (I.35) 39
- Antennal scape with either an outgrowth or a flange across it at the angle (I.36 & 37) 40

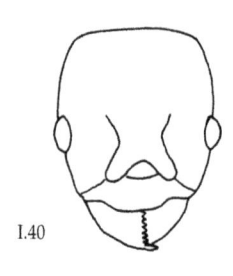

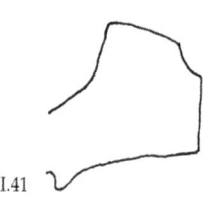

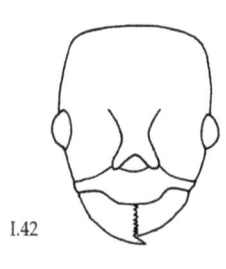

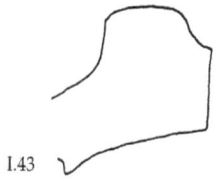

39 Upper surface of petiole curved into hind face
as in I.38; postpetiole almost spherical
Myrmica bessarabica
− Upper surface of petiole angled with hind face as in I.39; postpetiole high, rectangular in side view
Myrmica scabrinodis (pl. 8.2)

M. bessarabica and *M. scabrinodis* are very similar species. Although some specimens are readily identified, others are difficult because of the variable appearance of both species. Some *M. sabuleti* may also be confused here if the outgrowth on the scape is small, so consult an expert if in doubt.

40 Angle of antennal scape with an outgrowth (I.36)
Myrmica sabuleti
− Angle of antennal scape with a flange across it (I.37)   41

41 Frons about one third width of head (I.40); petiole node with an acute angle between the front and top faces (I.41)
Myrmica lobicornis
− Frons narrow, about one quarter width of head (I.42); petiole node with an obtuse angle between the front and top faces, the latter of which is a little domed (I.43)
Myrmica schencki

II Queen ants

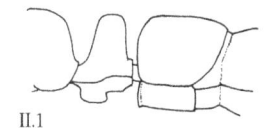

The characters used to identify queens are often similar to those of the workers of the same species.

1   Waist consisting of one segment, the petiole
(II.1, 2 & 3)   2
–   Waist consisting of two segments, the petiole and postpetiole (II.4)   Myrmicinae 24

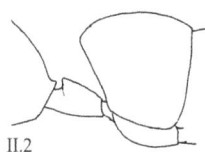

2   Gaster with a distinct constriction between the first and second segments (II.1); sting present
Ponerinae *Ponera coarctata* (pl. 10.1)
–   Gaster with no constriction between the first and second segments (II.2); sting absent   3

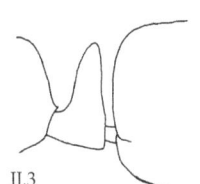

3   Petiole very small, overhung by the first segment of the gaster, not visible from above (II.2)
Dolichoderinae, genus *Tapinoma* 4
–   Petiole well developed, scale-like, clearly visible from above (II.3)   Formicinae 5

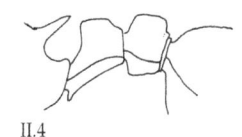

4   Notch in clypeus shallower, as in I.5; (ant generally smaller)   *Tapinoma ambiguum*
–   Notch in clypeus deeper, as in I.6; (ant generally larger)
Tapinoma erraticum (pl. 10.8)

These two species of *Tapinoma* are very similar and best separated by characters of the male

5   Opening of propodeal spiracle (breathing pore) elongate, on side face of propodeum (II.5); hind tibia with a double row of bristles present on the underside (in addition to those at the tip)   genus *Formica* 6
–   Opening of propodeal spiracle circular or nearly so, on curved surface between side and rear faces (II.6); hind tibia with no rows of bristles on the underside, although there are bristles at the tip (where there are also usually abundant, very short, soft hairs pressed flat against the tibia)   genus *Lasius* 16

6   Front margin of clypeus with a notch in the middle, as in I.9   *Formica sanguinea*
–   Front margin of clypeus without a notch in the middle   7

7   Hind margin of head strongly concave, as in I.10
Formica exsecta (pl. 10.9)
–   Hind margin of head not strongly concave, as in I.11   8

8   Thorax dark, the same colour all over   9
−   Thorax with reddish areas   11

9   Head and thorax strongly shining black; lower surface of
    head with a few long hairs   *Formica candida*
−   Head and thorax black, occasionally brownish-black, not
    strongly shining; lower surface of head without hairs   10

10 Row of erect hairs on top of pronotum extending round
    to wing bases; femora of middle legs with a few long
    hairs beneath   *Formica lemani*
−   Row of erect hairs on top of pronotum confined to the
    front portion; femora of middle legs without long hairs
                                                  Formica fusca

11 Frontal triangle finely rough, dull, not reflecting light   12

−   Frontal triangle reflecting light or shining, unsculptured
    or with a few very fine pits or with a crescent-shaped
    area of quite dense fine pits across it   13
    The wood ants – a difficult group. Consult an expert if in doubt.

12 Top of propodeum with a few long hairs; mesoscutum
    usually two-coloured   *Formica rufibarbis*
−   Top of propodeum without long hairs; mesoscutum
    usually uniformly dark, occasionally with reddish
    around wing bases   *Formica cunicularia*

13 Mesoscutum and scutellum both matt and dull
                                                 Formica pratensis
−   Mesoscutum more densely sculptured than the
    scutellum, which is shining   14

14 Erect hairs not present on compound eyes; no erect hairs
    on the hind margin of the head, as in I.12
                                                 Formica rufa (pl. 1.6)
−   Erect hairs present on compound eyes; erect hairs form a
    fringe (occipital fringe) on the hind margin of the head 15

15 Upper surface of petiole with long hairs; occipital fringe
    extending round head to compound eyes, as in I.13
                                                 Formica lugubris
−   Upper surface of petiole without long hairs; occipital
    fringe not extending as far as compound eyes, as in I.14
                                                 Formica aquilonia (pl. 6.4)

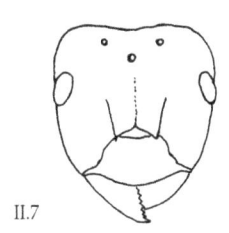

16   Colour strongly shining black (head II.7)
              Lasius fuliginosus (pl. 10.10)
  – Colour dark brown or yellowish brown, not strongly
     shining   17

17   Head width about the same as width of thorax (II.8); size
     distinctly smaller, about twice length of the workers of
     the same species   18
  – Thorax wider than head (II.9); size larger, about
     3–4 times the length of the workers of the same species 21

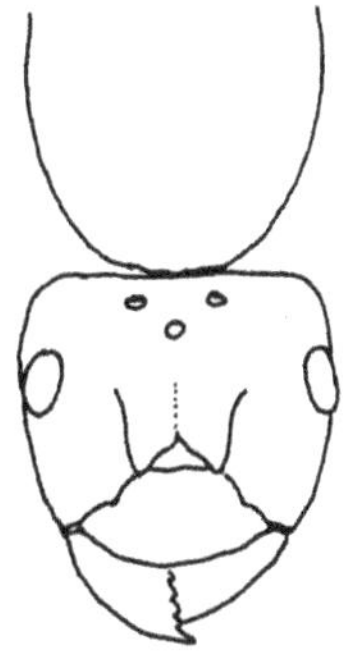

18   Hind tibiae without erect hairs; antennal scape without
     slanting hairs   *Lasius mixtus*
  – Hind tibiae with standing hairs; antennal scape with
     slanting hairs   19

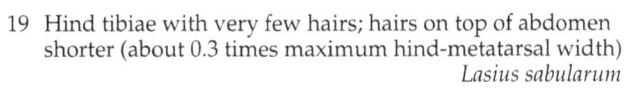

19   Hind tibiae with very few hairs; hairs on top of abdomen
     shorter (about 0.3 times maximum hind-metatarsal width)
                                                   Lasius sabularum

     L. sabularum, *L. meridionalis* and *L. umbratus* are very similar species -
     consult an expert if in doubt.

  – Hind tibiae with more hairs; hairs on top of abdomen
     longer (about 0.6 times maximum hind-metatarsal width)
                                                                          20

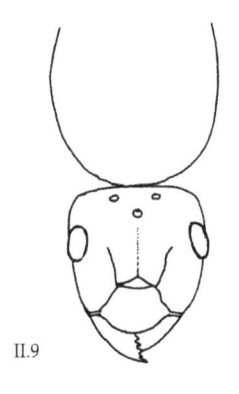

20   Scape somewhat flattened, with a narrow leading edge,
     at mid-point maximum width about twice minimum
     width; penultimate joint of antenna distinctly longer than
     wide   *Lasius meridionalis*
  – Scape almost circular in section, at mid-point maximum
     width about one and a half times minimum width;
     penultimate joint of antenna a little longer than wide
                                                   Lasius umbratus

21   Antennal scape and hind tibia with standing hairs
                                                       Lasius niger

     Seifert (1992) has shown that the species known as *Lasius niger*
     comprises two species in the UK, *L. niger* in the strict sense and
     L. platythorax. These species are very similar and in the queen, they can
     only be distinguished by very accurate measurement. *L. platythorax* is
     said to have a longer, flatter thorax. It occurs in woodland and wetland,
     and hence has a more northerly distribution.

  – Antennal scape and hind tibia without standing hairs 22

22   Wings not smoky at base   *Lasius alienus*
     Seifert (1992) recognises *L. psammophilus* as a species distinct from
     L. alienus. Consult an expert if a species identification is important.
  – Wings smoky, sometimes weakly so, at base   23

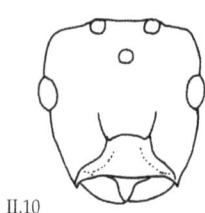

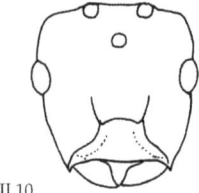

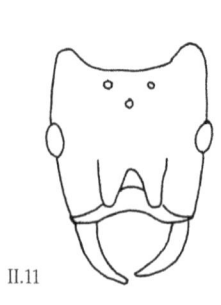

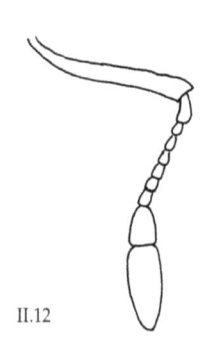

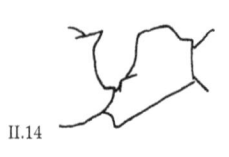

23 Frontal triangle clearly defined; compound eyes without short hairs between facets *Lasius brunneus*
— Frontal triangle not well defined; compound eyes with short hairs between the facets *Lasius flavus*

24 Mandibles with weak or indistinct teeth 25
— Mandibles with at least 5 normal teeth 26

25 Antennae with 10 or 11 segments; mandibles short, toothless apart from one tooth at the tip (II.10) *Anergates atratulus* (pl. 8.3, 10.6)
— Antennae with 12 segments; mandibles long, curved, pointed and toothless (II.11) *Strongylognathus testaceus* (pl. 10.4)

26 Antennae with 10 segments, the two at the tip forming a club (II.12); propodeum without any spines or teeth, as in I.21 *Solenopsis fugax*
— Antennae with 11 or 12 segments; propodeum armed with two teeth or spines, as in I.22 27

27 Lower surface of head with a pair of ridges along it, one each side, as in I.23; front margin of clypeus with two prominent teeth, sometimes with a small one in the middle, as in I.24 *Myrmecina graminicola* (pl. 10.3)
— Lower surface of head without a pair of ridges; front margin of clypeus without teeth 28

28 Sides of clypeus with their hind margins raised into a sharp, transverse ridge in front of antennal bases, as in I.25 *Tetramorium caespitum* (pl. 10.2)
— Sides of clypeus not raised into a transverse ridge 29

29 Middle of clypeus with two small ridges along it; petiole elongate in front (II.13) genus *Stenamma* 30
— Middle of clypeus without ridges; petiole not elongate in front (II.14) 31

30 Width of central smooth area of frons greater than one third the total distance between bases of antennae, as in I.26 (ant with shorter, stouter legs) *Stenamma debile*
— Width of central smooth area of frons less than one third the total distance between bases of antennae, as in I.27 (ant with longer, narrower legs) *Stenamma westwoodii*

These species of *Stenamma* are very similar and difficult to separate. Consult an expert if in doubt.

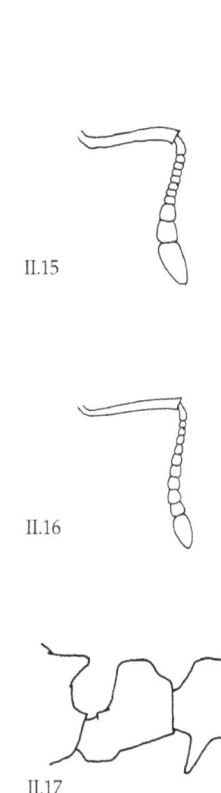

31 Three segments at tip of antenna as long as the remainder of the funiculus (II.15); mandibles with 5 or 6 teeth 32
–   Three segments at tip of antenna distinctly shorter than remainder of funiculus (II.16); mandibles with occasionally 6, usually 7 or more, teeth genus *Myrmica* 36

II.15

32 Postpetiole with a projection below (II.17); upper surfaces of head and thorax smooth and shining
    Formicoxenus nitidulus
–   Postpetiole without a projection below (II.18); upper surfaces of head and thorax sculptured
    genus *Leptothorax* 33

II.16

33 Antennae with 11 segments
    Leptothorax acervorum (pl. 10.7)
–   Antennae with 12 segments 34

34 Last segments of antenna the same colour as the rest of the antenna (reddish) *Leptothorax nylanderi*
–   Last segments of antenna darkened 35

II.17

35 Spines on propodeum long and curved; petiole node with a sharp angle on top *Leptothorax interruptus*
–   Spines on propodeum short and straight; petiole node a little flattened on top *Leptothorax tuberum*

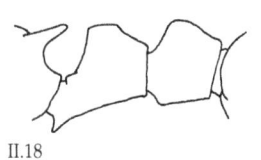

36 Postpetiole with a projection below; first dorsal plate on gaster without standing hairs *Myrmica karavajevi*

   M. karavajevi is a social parasite without a worker caste, and is found in nests of *M. sabuleti* and *M. scabrinodis*.

II.18

–   Postpetiole without a projection below; first dorsal plate on gaster with standing hairs 37

37 Postpetiole unusually broad when viewed from above, broader than high; a small species, approximately 4.5 mm (hairs of legs, antennae and body denser and more slanting, particularly noticeable on the antennal scapes; spines on propodeum shorter and thicker than in all the following *Myrmica* species except *M. rubra*)
    Myrmica hirsuta (pl. 3.5)

   A social parasite. Workers are rare. It is found in nests of *M. sabuleti*.

–   Postpetiole of normal breadth, that is higher than broad; larger species, usually 5.0 mm or more (hairs less dense and more erect; spines on propodeum usually longer and narrower) 38

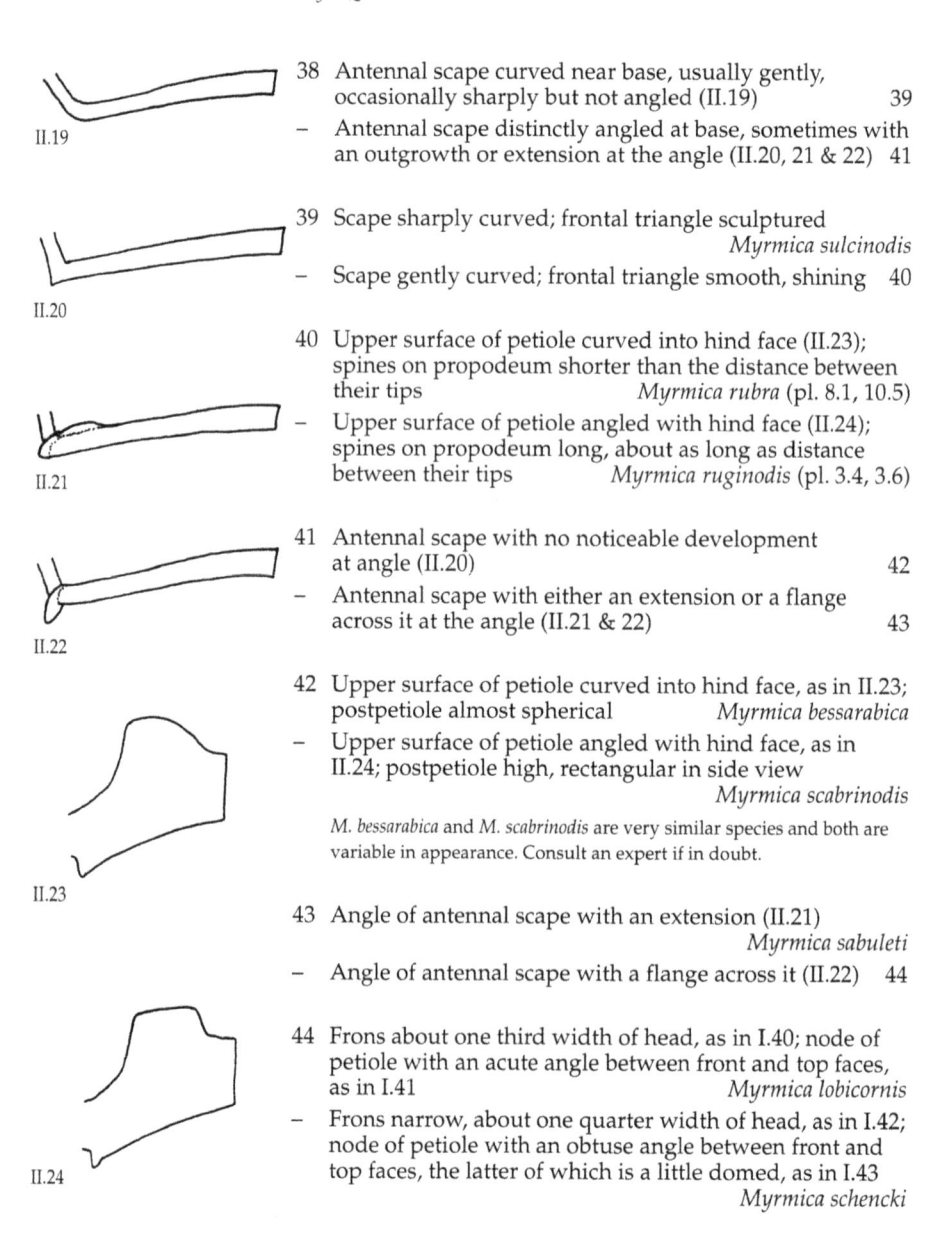

38 Antennal scape curved near base, usually gently, occasionally sharply but not angled (II.19)   39
 – Antennal scape distinctly angled at base, sometimes with an outgrowth or extension at the angle (II.20, 21 & 22)   41

39 Scape sharply curved; frontal triangle sculptured
Myrmica sulcinodis
 – Scape gently curved; frontal triangle smooth, shining   40

40 Upper surface of petiole curved into hind face (II.23); spines on propodeum shorter than the distance between their tips   *Myrmica rubra* (pl. 8.1, 10.5)
 – Upper surface of petiole angled with hind face (II.24); spines on propodeum long, about as long as distance between their tips   *Myrmica ruginodis* (pl. 3.4, 3.6)

41 Antennal scape with no noticeable development at angle (II.20)   42
 – Antennal scape with either an extension or a flange across it at the angle (II.21 & 22)   43

42 Upper surface of petiole curved into hind face, as in II.23; postpetiole almost spherical   *Myrmica bessarabica*
 – Upper surface of petiole angled with hind face, as in II.24; postpetiole high, rectangular in side view
Myrmica scabrinodis

M. bessarabica and *M. scabrinodis* are very similar species and both are variable in appearance. Consult an expert if in doubt.

43 Angle of antennal scape with an extension (II.21)
Myrmica sabuleti
 – Angle of antennal scape with a flange across it (II.22)   44

44 Frons about one third width of head, as in I.40; node of petiole with an acute angle between front and top faces, as in I.41   *Myrmica lobicornis*
 – Frons narrow, about one quarter width of head, as in I.42; node of petiole with an obtuse angle between front and top faces, the latter of which is a little domed, as in I.43
Myrmica schencki

III Male ants

1 Waist consisting of one segment, the petiole (III.1 & 2) 2
－ Waist consisting of two segments, the petiole and postpetiole (III.3) Myrmicinae 24

2 Gaster with a distinct constriction between the first and second segments (III.1); plate on underside of last segment of gaster elongated into a down-curved spine
Ponerinae *Ponera coarctata* (pl. 7.2)
－ Gaster with no constriction between the first and second segments (III.2); plate on underside of last segment of gaster not elongated 3

3 No erect hairs on top of thorax (III.4)
Dolichoderinae, genus *Tapinoma* 4
－ Erect hairs present on top of thorax (III.5) Formicinae 5

4 Head rounded behind; scape of antenna extending back well beyond hind margin of head; side lobes of subgenital plate (plate on underside of last segment of gaster) widely separated and narrow (III.6)
Tapinoma ambiguum
－ Head squarer behind; scape of antenna extending only a little beyond hind margin of head; side lobes of subgenital plate closer together and broader (III.7)
Tapinoma erraticum (pl. 10.8)

5 Propodeal spiracle (breathing pore) elongate, slit-like, on side of propodeum, as in II.5 genus *Formica* 6
－ Propodeal spiracle circular or nearly so, on curved surface where side rounds into hind face, as in II.6
genus *Lasius* 16

6 Front margin of clypeus with a notch in the middle (III.8)
Formica sanguinea
－ Front margin of clypeus smoothly rounded (III.9 & 10) 7

7 Hind margin of head strongly concave (III.9)
Formica exsecta
－ Hind margin of head not strongly concave (III.10) 8

8 Compound eyes with erect hairs 9
－ Compound eyes without erect hairs 12

Key III Male ants

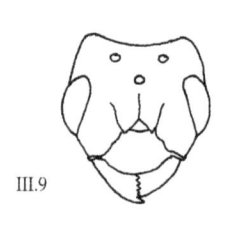

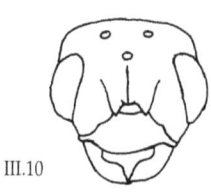

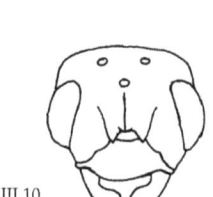

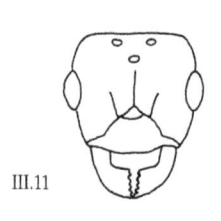

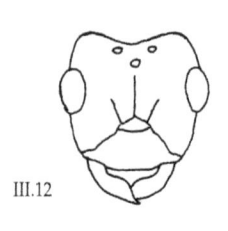

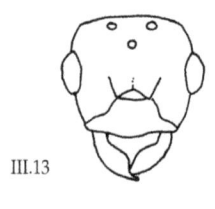

III.9

III.10

III.11

III.12

III.13

9 Side of head between compound eye and clypeus without long erect hairs   *Formica rufa*
- Side of head between compound eye and clypeus with long erect hairs   10

10 Gaster dull, not shining; erect hairs arising over entire surface of plate on upperside of second segment of gaster   *Formica pratensis*
- Gaster shining; erect hairs arising mainly from front portion of plate on upperside of second segment of gaster   11

11 Side of head between eye and clypeus with only two or three erect hairs; gaster with only a few erect hairs   *Formica aquilonia*
- Side of head between eye and clypeus with numerous erect hairs; gaster with numerous, scattered, erect hairs   *Formica lugubris* (pl. 1.1)

12 Underside of head with one or two long hairs; gaster shining   *Formica candida*

    The males of this and the following four *Formica* species are difficult to separate. Consult an expert if in doubt.

- Underside of head without hairs; gaster dull or scarcely shining   13

13 Petiole with a fringe of very short hairs   *Formica fusca*
- Petiole with short hairs and also with long erect hairs   14

14 Scutellum and gaster a little shining   *Formica lemani*
- Scutellum and gaster dull   15

15 Femora mainly dark   *Formica rufibarbis*
- Femora mainly pale   *Formica cunicularia*

16 Mandible with two or more teeth   17
- Mandible with only one tooth, at the tip (III.12 & 13)   20

17 Mandible with two teeth, one at the tip and one just before the tip   *Lasius sabularum*

    L. sabularum, *L. mixtus*, *L. meridionalis* and *L. umbratus* are very similar and difficult to separate. Consult an expert if in doubt.

- Mandibles with many teeth on the broad cutting edge (III.11)   18

18 Tibiae and scape of antenna without erect hairs
                                                    Lasius mixtus
 – Tibiae and scape of antenna with erect hairs   19

19 Frons shining, with very little microsculpture
                                                Lasius meridionalis
 – Frons dull, microsculptured   *Lasius umbratus*

20 Colour shining black; hind margin of head strongly
    concave (III.12)   *Lasius fuliginosus*
 – Colour brownish-black; hind margin of head
    not strongly concave (III.13)   21

21 Tibiae and scape of antenna with erect hairs *Lasius niger*
    Seifert (1992) has shown that the species known as *Lasius niger* comprises two species in the UK: true *L. niger* and the very similar *L. platythorax*. There are no known characters to separate the males.
 – Tibiae and scape of antenna without erect hairs   22

22 Recurrent vein of forewing always absent (III.14), there
    being no discoidal cell in the middle of the wing; furrow
    on frons indistinct (III.13)   *Lasius flavus* (pl. 9.3)
 – Recurrent vein of forewing usually present (III.15),
    defining a discoidal cell in the middle of the wing;
    furrow on frons distinct   23

III.14

23 Sides of head behind compound eyes usually without
    hairs; wings smoky near base   *Lasius brunneus*
 – Sides of head behind eyes with hairs; wings clear
                                                Lasius alienus

III.15

Seifert (1992) recognises *L. psammophilus* as a species distinct from *L. alienus*. Consult an expert if a species identification is important.

24 Wingless   25
 – Winged   26

25 Antennae with 10 or 11 segments; ant looks like a pupa
                                              Anergates atratulus
 – Antennae with 12 segments; ant very worker-like in
    appearance   *Formicoxenus nitidulus*

III.16

26 Antennae with 10 segments and elongate third segment
    (III.16)   27
 – Antennae with 12 or 13 segments   28

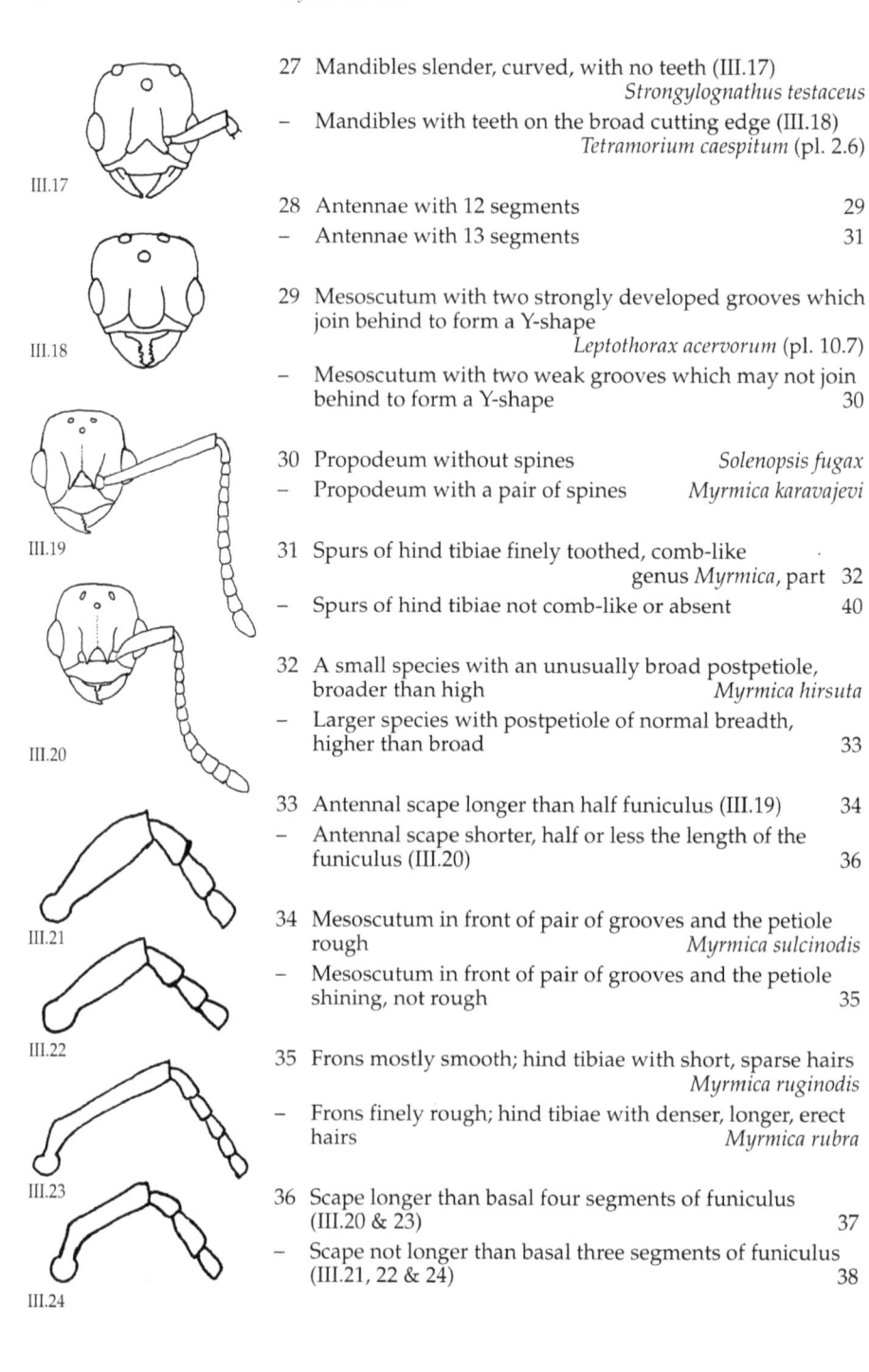

27 Mandibles slender, curved, with no teeth (III.17)
Strongylognathus testaceus
— Mandibles with teeth on the broad cutting edge (III.18)
Tetramorium caespitum (pl. 2.6)

28 Antennae with 12 segments 29
— Antennae with 13 segments 31

29 Mesoscutum with two strongly developed grooves which join behind to form a Y-shape
Leptothorax acervorum (pl. 10.7)
— Mesoscutum with two weak grooves which may not join behind to form a Y-shape 30

30 Propodeum without spines *Solenopsis fugax*
— Propodeum with a pair of spines *Myrmica karavajevi*

31 Spurs of hind tibiae finely toothed, comb-like
genus *Myrmica*, part 32
— Spurs of hind tibiae not comb-like or absent 40

32 A small species with an unusually broad postpetiole, broader than high *Myrmica hirsuta*
— Larger species with postpetiole of normal breadth, higher than broad 33

33 Antennal scape longer than half funiculus (III.19) 34
— Antennal scape shorter, half or less the length of the funiculus (III.20) 36

34 Mesoscutum in front of pair of grooves and the petiole rough *Myrmica sulcinodis*
— Mesoscutum in front of pair of grooves and the petiole shining, not rough 35

35 Frons mostly smooth; hind tibiae with short, sparse hairs
Myrmica ruginodis
— Frons finely rough; hind tibiae with denser, longer, erect hairs *Myrmica rubra*

36 Scape longer than basal four segments of funiculus (III.20 & 23) 37
— Scape not longer than basal three segments of funiculus (III.21, 22 & 24) 38

III.25

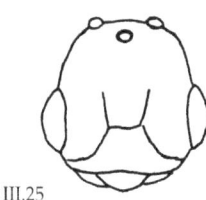

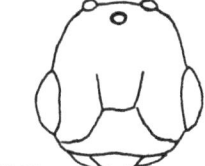

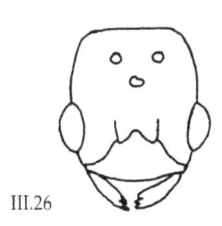

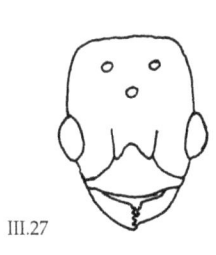

III.26

III.27

37 Scape sharply angled at base, longer than basal five
segments of funiculus (III.23)　　*Myrmica lobicornis*
- Scape bent, thick, longer than four but shorter than basal
five segments of funiculus (III.20)　　*Myrmica sabuleti*

38 Scape thick (III.21), hind tarsal hairs dense and long
　　Myrmica scabrinodis
- Scape thinner (III.22 & 24), hind tarsal hairs shorter and
less dense　　39

39 Frons impressed just forward of middle ocellus; second
segment of funiculus almost twice length of first (III.24)
　　Myrmica schencki
- Frons not impressed forward of middle ocellus; second
segment of funiculus about same length as first (III.22)
　　Myrmica bessarabica

40 Mandibles small, not functional, toothless (III.25)
　　Myrmecina graminicola
- Mandibles of normal size, always toothed　　41

41 Petiole elongate in front, as in II.13; propodeum with a
small pair of teeth　　genus *Stenamma* 42
- Petiole not elongate in front, as in II.14; propodeum not
armed with a small pair of teeth
　　genus *Leptothorax* (part) 43

42 Mandible with three teeth, appearing narrow with no
marked cutting edge (III.26)　　*Stenamma debile*
- Mandible with five teeth on the broad cutting edge
(III.27)　　*Stenamma westwoodii*

43 Front portion of mesoscutum between pair of grooves
smooth and shining　　*Leptothorax nylanderi*
- Front portion of mesoscutum between pair of grooves
sculptured　　44

44 Scutellum smooth on top　　*Leptothorax tuberum* (pl. 5.1)
- Scutellum sculptured on top apart from a narrow strip in
the middle　　*Leptothorax interruptus*

Quick-check field key to common British ants

Beware: this key is highly simplified and includes only the very commonest species. For serious identification, use keys I–III.

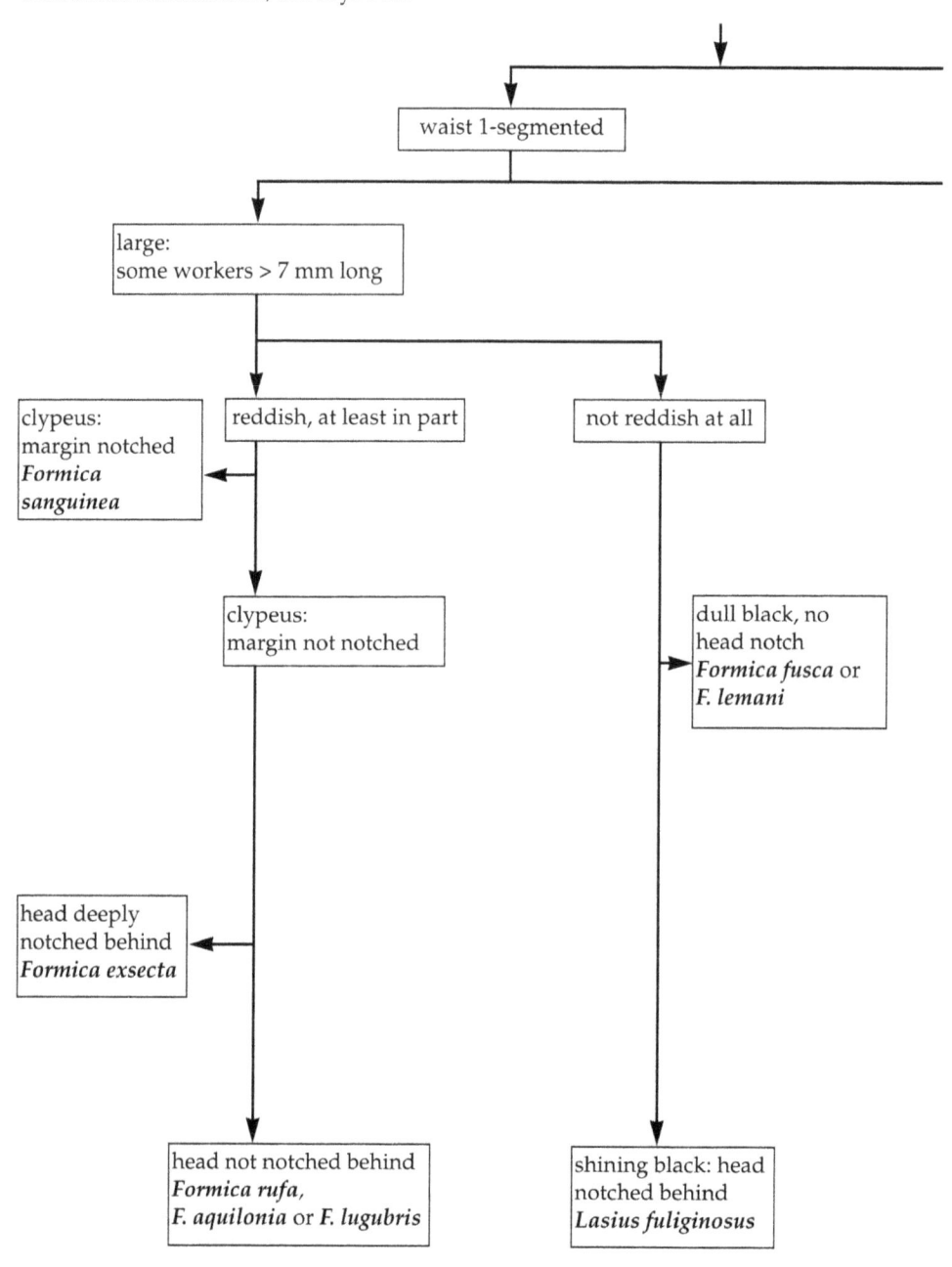

```
                              ↓
              ┌───────────────┴───────────────┐
              │      waist 1-segmented        │
              └───────────────┬───────────────┘
                              ↓
              ┌───────────────────────────────┐
              │ large:                        │
              │ some workers > 7 mm long      │
              └───────┬───────────────┬───────┘
                      ↓               ↓
       ┌──────────────────────┐  ┌──────────────────┐
       │ reddish, at least in │  │ not reddish at   │
       │ part                 │  │ all              │
       └──────────┬───────────┘  └────────┬─────────┘
                  │                       │
   ┌──────────────┤                       ├──────────────────┐
   │ clypeus:     │                       │ dull black, no   │
   │ margin       │←                      │ head notch       │
   │ notched      │                       │ Formica fusca or │
   │ Formica      │                       │ F. lemani        │
   │ sanguinea    │                       └──────────────────┘
   └──────────────┘
                  ↓
       ┌──────────────────────┐
       │ clypeus:             │
       │ margin not notched   │
       └──────────┬───────────┘
                  │
   ┌──────────────┤
   │ head deeply  │
   │ notched      │←
   │ behind       │
   │ Formica      │
   │ exsecta      │
   └──────────────┘
                  ↓                       ↓
       ┌──────────────────────┐  ┌──────────────────────┐
       │ head not notched     │  │ shining black: head  │
       │ behind               │  │ notched behind       │
       │ Formica rufa,        │  │ Lasius fuliginosus   │
       │ F. aquilonia or      │  └──────────────────────┘
       │ F. lugubris          │
       └──────────────────────┘
```

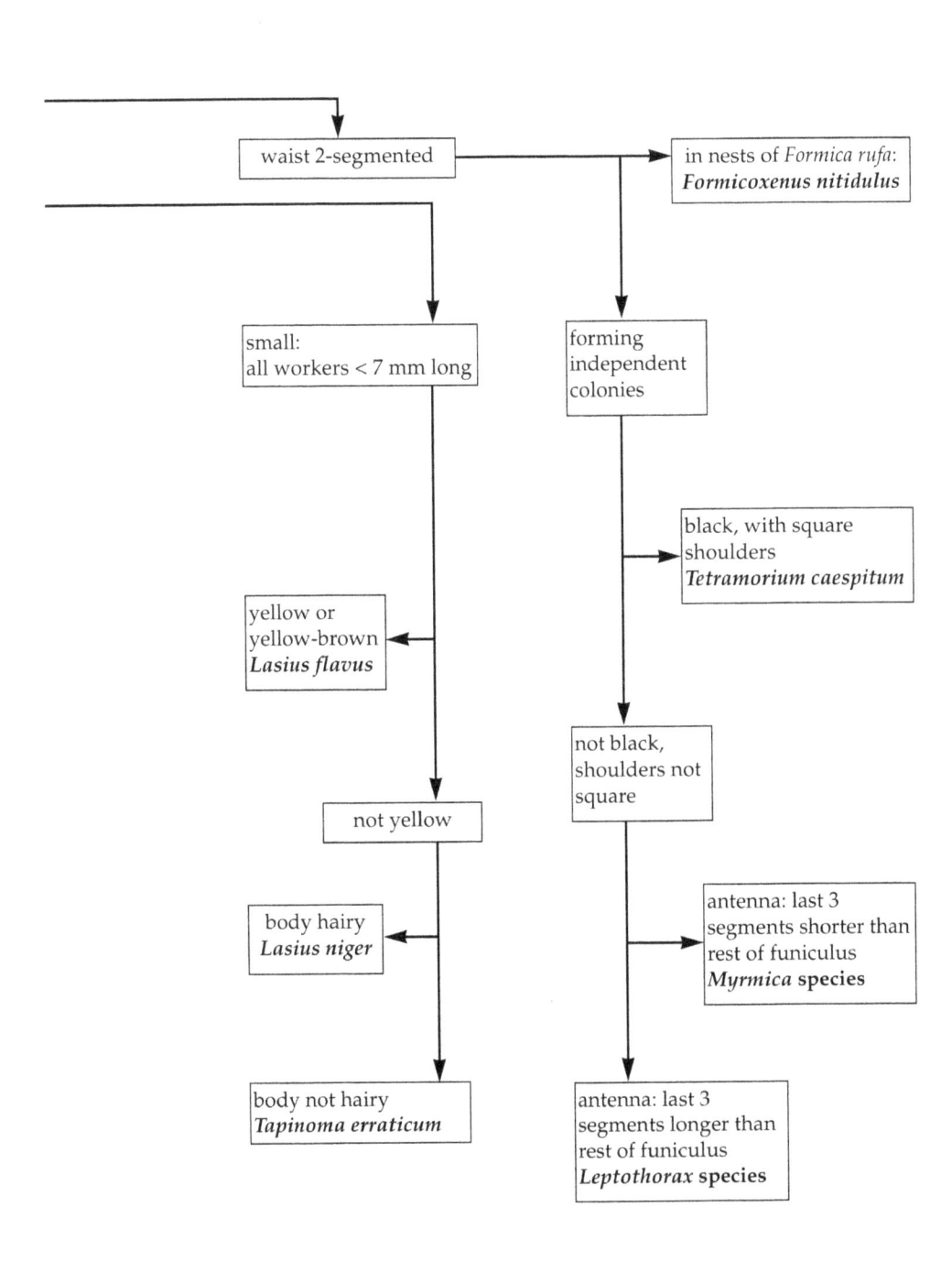

```
                    ┌──────────────────┐       ┌──────────────────────┐
────────────────────│ waist 2-segmented │──────▶│ in nests of *Formica rufa*: |
                    └──────────────────┘       │ ***Formicoxenus nitidulus*** │
                             │                 └──────────────────────┘
─────────────────────────────┤
                             ▼
```

- **waist 2-segmented**
  - → in nests of *Formica rufa*: ***Formicoxenus nitidulus***
  - forming independent colonies
    - black, with square shoulders ***Tetramorium caespitum***
    - not black, shoulders not square
      - antenna: last 3 segments shorter than rest of funiculus ***Myrmica* species**
      - antenna: last 3 segments longer than rest of funiculus ***Leptothorax* species**
  - small: all workers < 7 mm long
    - yellow or yellow-brown ***Lasius flavus***
    - not yellow
      - body hairy ***Lasius niger***
      - body not hairy ***Tapinoma erraticum***

Notes on the commoner British species

About fifty species of ants can be regarded as British, not including the so-called tramp species which live in hothouses of various kinds. Many of our truly British species are rare and very unlikely to be seen by the beginner. The following notes describe the most common and widely distributed of the British ants. The keys in this chapter include all the species likely to be encountered in Britain.

Ponera coarctata
A primitive ant forming small colonies beneath stones. It is found in warm places around the south coast, and there are some records from the Thames Valley.

Myrmica species
Myrmica is the largest ant genus in Britain, with ten currently recognised species.

M. rubra
Found throughout Britain, but becoming less common in the north. It forms small nests of a hundred or so individuals under stones and in tree stumps.

M. ruginodis
A very common species, found throughout Britain, and rather more common than *M. rubra* in the north. Colonies have a few hundred workers.

M. scabrinodis
Found throughout Britain, and, like *M. rubra*, less common in the north. Workers tend to be rather smaller than the previous two species.

M. sabuleti
Less common than the three species so far mentioned, but still distributed throughout Britain, again tailing off in the north.

M. sulcinodis
Rather local in distribution and generally found on heaths in the uplands, although colonies exist on lowland heaths in the south. Colonies are small with about 150–200 workers. The species is almost absent from the midlands and most of Wales.

M. lobicornis
There are scattered records of this species throughout Britain but is never common.

M. schencki
Found only in the south midlands and sites in the south, and in Ireland. Nests are likely to be found only in warm, sheltered places.

Formicoxenus nitidulus
Although not common, this is an interesting species because it spends all its life nesting in colonies of the much larger *F. rufa* and its allies. It is found in most places where wood ants live, but is never as common as its hosts.

Leptothorax acervorum
Found throughout Britain, this small ant is easily overlooked, especially in the south where it nests in stumps and under bark. In the north it is found under stones or in peat.

Other *Leptothorax* species
The other three British species of *Leptothorax* (excluding one found only in the Channel Islands) are very southern in distribution and found in only a few sites.

Strongylognathus testaceus and *Anergates atratulus*
These are both parasites found only with *Tetramorium*. *Anergates* has no workers.

Tetramorium caespitum
This species tends to be coastal although it is found more commonly inland in the south. It extends northwards to the Isle of Man in the west, but only as far as East Anglia in the east, with an odd record in Northumberland. It forms large colonies, typically in heaths or sand on the coast.

Solenopsis fugax
This species has very small yellow workers, but the sexual forms are much larger. The nests are typically found under stones, particularly deeply buried ones, and they are often associated with those of *Lasius flavus* and *Formica* species. *S. fugax* has been found mainly in coastal sites in the southern counties and in the Channel Islands.

Stenamma westwoodii
This is a small and easily overlooked species nesting under stones in small colonies of about 100 workers. It is found throughout the south. All old records will need re-assessment as it is now known that there are two species in Britain (p. 44).

Myrmecina graminicola
This is a very dark species, found scattered throughout most of the south of England and recorded once or twice in Wales.

Tapinoma erraticum
This species is only found in the extreme south of Britain, including Dorset, Hampshire, Surrey and Kent. It nests in heathland.

Formica rufa
This species has large, reddish workers. It builds thatched mounds in woodland, mainly in less shady areas, clearings and woodland edges. The colonies consist of a few hundred thousand workers which tend to forage along well-defined trails from the nest. Some workers will forage over the general woodland floor, while others stream up trees to catch prey and tend aphids. It is southerly in distribution, extending into northern Cumbria in isolated sites, but it is not found in Scotland or Ireland.

F. lugubris
Very similar to *F. rufa*, *F. lugubris* tends to make smaller nests and have more inter-nest links than *F. rufa*. It is more northerly in distribution than *F. rufa*, but the two species overlap in North Wales and northern counties of England. *F. lugubris* is also found in parts of Ireland and the Scottish Highlands.

F. aquilonia
This species is similar to *F. rufa* and *F. lugubris*, building thatched mound nests. It is found commonly in the Highlands of Scotland, where it overlaps with *F. lugubris*, and there are one or two records from Ireland. It is not found in England.

F. pratensis
Probably now extinct; last seen in Dorset.

F. exsecta
Found only in the Bournemouth area, in parts of Devon and in the Highlands of Scotland. The nests are smaller than those of *F. rufa* and its allies.

F. sanguinea
Rather bigger and redder that the other red *Formica* species, *F. sanguinea* is interesting because it is a slavemaker. It raids nests of other species, particularly *F. fusca* and *F. lemani*, and carries their pupae back to its own nest. They are then integrated into the *F. sanguinea* nest, where they emerge to become functionally *F. sanguinea* workers. Mixed nests are thus formed. The species is found in restricted parts of south central Britain, and in Dorset, South Wales and the Scottish Highlands.

F. fusca and *F. lemani*
These two species are very similar and difficult to separate. More than one specimen should be examined from each colony as there is much variation in the key characters of hairiness of legs and thorax. The two are similar in their ecology, with average-sized nests, often under stones. They are very active in hot weather. *F. fusca* is southerly in distribution, extending up to north Lancashire, with a few scattered sites in West Scotland. *F. lemani* is more northerly and commonest in Scotland. It replaces *F. fusca* in most parts

of that country and in the north west of England. The two species overlap in the midlands and throughout Wales and the south west. In Ireland, both species are found, but *F. lemani* is the commoner.

F. cunicularia
This is a reddish species without the thatched mounds of the true wood ants. It is quite common around southern England, but has not been found north of Cambridgeshire.

F. rufibarbis
This species is rare. It has been found in a few sites in the south and in the Scillies.

F. candida
Rare, found only in Dorset and with one record from Wales.

Lasius fuliginosus
This is one of our most striking ants. The workers are large and shining black, and have a very characteristic heart-shaped head. It is southerly in distribution, but found as far north as the Isle of Man in the west. It extends only to East Anglia in the east. It nests in old tree stumps, making a papery substance called carton from chewed wood and saliva. It forages along distinct trails in the manner of the wood ants.

Lasius niger
L. niger is a common species and perhaps the one most often seen by the casual observer. It is frequently found under stones (including paving stones) in towns. Another reason for its familiarity is its tendency to enter houses.

L. alienus
This is very common on the coast in the south.

L. brunneus
Found only in the home counties, this species nests in trees.

L. flavus
Better known from its nests than its workers, this is the famous mound-building ant. The large, vegetation-covered, earthen mounds can be found in their hundreds in undisturbed grassland, but sometimes mounds are not built at all. It is found throughout Britain and is very common.

L. umbratus
Found throughout most of England, but commoner in the south, this species nests in logs and tree stumps and occasionally under stones.

L. mixtus
This is another rather southern species, but there are records from the Highlands of Scotland.

4 Techniques

Finding, collecting and preserving ants

Fig. 19. A pitfall trap set up correctly.

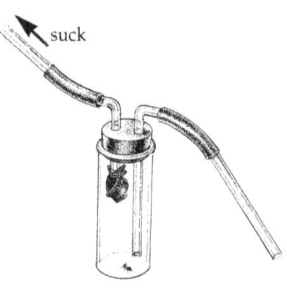

Fig. 20. A pooter for collecting ants.

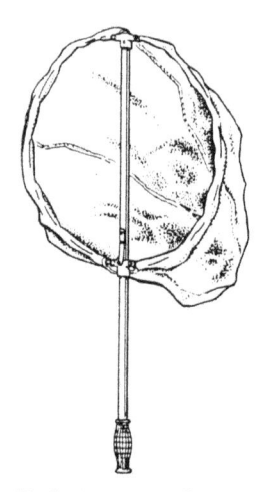

Fig. 21. A sweep net (from Majerus and Kearns, 1989).

Ants can be found almost everywhere, even in urban areas and gardens. Many species forage on the ground and if the weather is not too cold they can usually be found by watching an area of ground closely for a few minutes. Another favoured technique for the collection of such species is the pitfall trap. This consists of a glass jar or plastic cup, sunk into a hole in the ground to its very rim (fig. 19). Surface running species fall in. If live specimens are needed the trap should be used dry, if necessary with the inner surfaces smeared with a suspension of PTFE (polytetra-fluoroethene, Fluon) to prevent ants from escaping, and it should be checked frequently. If dead, preserved specimens will suffice, escapes can be prevented by charging the trap to a depth of about 2 cm with liquid, preferably 70% alcohol with a drop of detergent (washing-up liquid) per litre to reduce surface tension. Specimens caught in this way will not be so good for the museum drawer as those caught alive and killed in one of the ways described below. When left in the liquid the dead ant's body swells and becomes distorted, and when dried it is rigid. The specimen can be partially relaxed by leaving it for about 24 hours in a humid atmosphere in a tin or rustless container with a tight-fitting lid. In the bottom is absorbent material such as paper tissue moistened with water and covered with a layer of cork or balsa wood. Water must not run onto the specimen, so the lid is lined with blotting paper to absorb any moisture that condenses there. A little naphthalene or paradichlorobenzene in the tin will prevent mould from forming. Alternatively, chopped laurel leaves can be used as a relaxing agent as they provide moisture and also give off hydrogen cyanide which inhibits mould growth, but these must be treated with great care because hydrogen cyanide is very poisonous. Some of the firms mentioned on p. 74 supply a relaxing fluid that can be painted directly onto stiff insects. Freshly-killed specimens are always to be preferred for making up a serious collection (see below).

Ants must be handled carefully to avoid damage. Stiff narrow forceps are usually adequate, but fine watchmaker's forceps are best for very small specimens (Hölldobler & Wilson, 1990). Alternatively, specimens can be sucked up into a pooter (aspirator) (fig. 20). This method is inconvenient for handling some of the larger wood ants, which produce significant quantities of formic (methanoic) acid, causing the collector discomfort when sucking at the pooter. The discomfort can be reduced by lining the bottom of the pooter tube with absorbent material such as plaster of Paris, or, in the field in an emergency, with a little damp soil. To collect ants from herbaceous vegetation and low branches, a sweep net is very effective. This consists of a long, stout calico bag supported on a (usually) D-shaped metal frame (fig. 21). It is dragged or 'swept' vigorously through the vegetation and the end is flicked over the rim to close it. The contents are inspected carefully with pooter, specimen tubes and forceps handy so that specimens can be transferred quickly. To sample more widely in the tree, beating may be used (fig. 22). A large, pale sheet is laid out below selected branches and then these are struck firmly with a stick. Many crawling insects are dislodged and fall onto the sheet, from which they can easily be picked up with forceps, a pooter or a specimen tube.

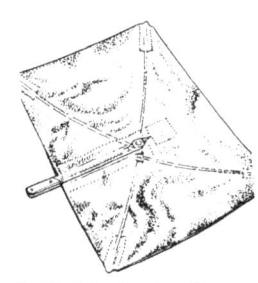

Fig. 22. A beating tray (from Majerus & Kearns, 1989).

These techniques work well for most species of ant in Britain, but it is harder to find the small species that nest in rotting logs and twigs and do not go far afield to forage, those living mainly underground and those inhabiting the nests of other species. It is often worth searching for the nest rather then the ants themselves. Nests are often the first clue to the presence of the yellow ant *Lasius flavus*. These ants forage mainly underground and workers are rarely seen at the surface, but the nests are very conspicuous (p. 3). Another good way of searching out nests of terrestrial species is to look underneath stones and logs lying on the ground. Workers, males queens and immature stages can often be sampled from such sites. The stones or logs should be replaced carefully, in exactly the original position, as soon as possible. Some species nest in rotting logs on the ground and the rotten wood is worth searching, preferably in a white tray such as a photographer's developing dish.

A combination of the methods outlined above, applied with some care in a particular habitat, should give a complete species list.

A reference collection of ants is sometimes required. Specimens are best killed in ethyl acetate vapour, but care must be taken not to inhale this or get it on the skin. A glass specimen tube should be used because ethyl acetate attacks many plastics. A few drops of ethyl acetate are absorbed onto plaster of Paris or strips of filter paper. The ants should be left in this killing bottle for an hour or so to ensure that they are quite dead. Ethyl acetate leaves the dead insects in a relaxed state, ideal for mounting neatly. They are best mounted straight away on a piece of stiff white card, big enough to hold the insect with its legs (and wings if present) fully spread out, leaving a small margin all around the edge with extra space for the pin at the back (fig. 23).

Fig. 23. An ant mounted on card in display fashion.

With a little practice, mounting is quite easy. The best mounting glue is gum tragacanth, which can be bought from a chemist. Only small amounts are needed. A level teaspoonful is placed in a 7.5 cm x 2.5 cm specimen tube. Water is then added, a little at a time, with continual stirring, until the tube is about half full. This will take some time, but it makes enough glue to mount thousands of ants, and this will keep for a long time if a few drops of alcohol are stirred in as a preservative. Alternatively, entomological suppliers sell ready made gum for mounting beetles, as 'Coleoptera mounting agent'.

Every experienced ant worker has his own idiosyncratic way of mounting ants. One method is as follows. The freshly-killed ant is placed upside down on the work surface with the ant's head facing away from the worker. A fine paint brush, dipped in gum, is applied, gently, to the ant's lower side. This will pick the ant up. Every effort should be made to ensure that the legs are not caught up in the gum. The brush is gently twisted to bring the animal right way up and the ant is lowered onto the prepared card. The brush is then slipped out backwards, leaving the ant on the card. The legs (and wings if present) may need some final arrangement and this is best done with a fine mounted needle, the fine brush, or, in the case of very small specimens, a very fine pin (a minuten) held firmly in curved or entomological forceps. Alternatively, the ant may be mounted on its side. This allows access to some characters not easily seen in ants mounted ventral side down.

An entomological pin can now be pushed firmly through the wide rear margin of the card, so that the card is about 1 cm below the top of the pin. This leaves the rest of the shaft of the pin, below the card, for labels. It is customary to use two labels. The upper one shows essential details: date of capture, locality, method and name of collector. The lower label shows the Latin name of the ant.

An alternative way to mount ants is to 'point' them. The ant is glued to the tip of a triangle of stiff card or white paper. This method permits examination of the underside of the specimen, which may occasionally be necessary. For this reason, the glue should touch as little as possible. This technique takes much less time, and is preferred by workers mounting many specimens for research. It does not produce such an attractive finished product and is not recommended for display purposes (fig. 24).

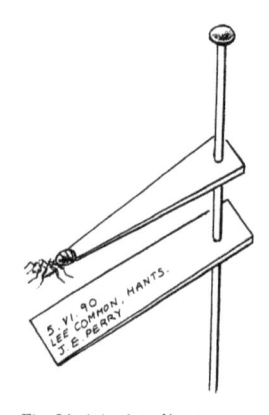

Fig. 24. A 'pointed' ant.

The whole collection is best kept in a proper storage box lined with cork or expanded polyethylene foam (plastazote) (suppliers p. 74). Quite good storage boxes can be made much more cheaply from any close-fitting box, deep enough to accommodate the pins and with a firmly-attached floor of cork, plastazote, or even polystyrene ceiling tiles. (Beware: ethyl acetate vapour destroys polystyrene.)

The minute amount of soft tissue in the ant's body will dry and shrink leaving the hard exoskeleton which, with care, will survive for decades or even centuries. The biggest problems are damp (which encourages mould growth) and attack by scavenging insects and mites. Mould can be avoided by storing the collection in a dry place. To prevent insect damage, the atmosphere inside the box is kept saturated with the vapour of naphthalene or paradichlorobenzene. Specially made insect boxes have a compartment for these crystals, but a home-made box will need some provision. A cardboard pepper pot (as sold in supermarkets) can be cut short and glued in the lid of the box. The pot can be filled with crystals and the plastic top with holes will allow vapour to diffuse into the box. Alternatively, the crystals can be enclosed in a twist of fine muslin, pinned into the box. They must not be allowed to escape, or they will damage specimens when the box is moved.

Larger numbers of specimens are best stored in 70% alcohol in small tubes. This may be necessary, for example, if one is collecting a series to show variation in worker ants from the same nest. Hölldobler & Wilson (1990) recommend that 20 specimens of each caste are collected, if possible, together with about 20 larvae, and are stored in small vials (55 mm x 8 mm). These are easily carried in the field and specimens can be put in immediately. Like dried specimens, those preserved in fluid must be properly labelled with all relevant data. An unlabelled specimen is valueless. The labels should be written in pencil and put inside the vial. Alcohol tends to evaporate eventually, even in a well-sealed container. To prevent this, the vials may be stored submerged in 70% alcohol in a large wide-mouthed jar with a well-fitting lid.

Culturing ants

Rearing ants indoors is easy, useful and interesting. By keeping colonies in 'formicaria' it is possible to extend the period during which research can be carried out in temperate latitudes, where ants out of doors hibernate in winter.

Ant culture has a long history. An early formicarium was described by White (1895). Many modifications and refinements have been made since then, but the basic principles remain the same. The ants must be enclosed to prevent escape, the founding colony must possess a functional queen, there must be provision for feeding the ants, the nest must remain moist for as long as possible and the ants must have access to a dark region of the formicarium.

There are effectively two types of nest, at least for the smaller species. Lubbock describes a glass observation nest in which the ants are confined to a very thin layer of soil between two horizontal glass plates, held apart with slips of wood, one of which is loose and slightly short to leave a door (fig. 25). The distance between the two plates should be adjusted to allow the ants to be seen when the assembly is uncovered, and will be somewhere between 2 mm and 6 mm for British species. To allow the ants to be in the dark, nests of this kind are kept covered when not under observation. Lubbock prevented the ants from escaping by placing the whole arrangement on a large wooden tray with a groove cut all around and filled with water to form a moat. To establish ants in such a formicarium, a suitable colony is dug up and placed over a prepared glass nest. As the original soil of the mound dries over a period of time, the ants vacate it and enter the moister soil between the two plates. The dry soil can be removed gradually leaving all the ants between the glass plates.

In contrast, 'Janet nests' contain no soil at all (fig. 26). Instead of being constructed by the ants in soil the chambers and galleries are pre-formed in some suitable

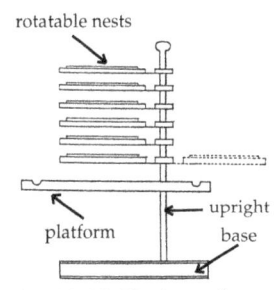

Fig. 25. A Lubbock nest (from Skaife, 1961).

Fig. 26. A Janet nest (from Skaife, 1961).

material. In its original form, this nest consists of four horizontal chambers of plaster of Paris. Three are interconnected by galleries. The fourth acts as a water trough keeping the others moist by capillary action through the porous plaster. The chamber furthest from the water trough is the driest. It is kept in the light and constitutes the food chamber. The other two chambers are kept covered when not being observed. The whole is covered with a glass sheet with a hole into each chamber and a second, removable glass sheet covering the holes.

Many modifications of these two basic types have been described and used over the last century or so (for example Wheeler, 1910; Donisthorpe, 1927; Skaife, 1961; Gösswald, 1985; Hölldobler & Wilson, 1990) and there is still room for improvisation in this area. Modern materials and artefacts surely provide opportunities for the development of quite new types of nest. Alternatively, it is possible to purchase an 'ant town' (supplier p. 74).

The types of nest described here are satisfactory for most British species, but not for very large colonies of large species, such as the wood ants. These can be housed in a large wooden or metal box. The biggest problem is keeping them in. Escape can be prevented by standing the whole box in a trough of water or smearing a layer of fruit-tree banding grease around the rim, but many of the workers will drown or become stuck and die. Also, the grease may become bridged by dead bodies, across which other workers may escape. Renewal of the grease at intervals will solve this problem. Again, there is scope for designing new ways of keeping these larger species.

The food provided for the ants will vary according to species, although Ettershank (1967) has invented a diet that would suit most ants. Some of his components are difficult to obtain, so a version using everyday materials is quoted here. This was invented by Bhatkar & Whitcomb (1970) and is used regularly by, for example, Hölldobler & Wilson.

Ingredients

1 egg
62 ml honey
1 g vitamins

*1 g minerals and salts
5 g agar
500 ml water

Dissolve the agar by adding it gradually to 250 ml boiling water, stirring constantly. Allow to cool. Mix 250 ml water, honey, vitamins, minerals and egg until smooth, using a liquidiser. Add the cooled agar to this with constant stirring. Pour this final mixture into shallow dishes such as Petri dishes. Store in a refrigerator.

This diet should be supplemented with such things as fragments of freshly killed mealworms, cockroaches or crickets, or whole fruit flies.

* The brand specified by Bhatkar & Whitcomb is not available in Britain, but Seven Seas multivitamin and mineral capsules (address p. 74) provide most of the necessary substances. Some experimentation may be needed.

More simply, colonies will often survive quite well on sugar water or honey water supplemented with insect fragments. The water is best supplied in a tube with a cotton wool wick, at which the ants come and drink.

Once a colony of ants is established in an observation nest, there is much scope for detailed study. Behaviour during chamber and gallery building can be observed in a Lubbock type nest. Janet and Lubbock nests are both suitable for observations on brood and queen care and food exchange behaviour within the nest. In observation nests with foraging arenas, foraging behaviour, food finding and recruitment can be investigated. Potted plants infested with aphids can be placed in larger arenas, providing an opportunity to gather information on tending behaviour and honeydew intake.

Throughout all such work it must be borne in mind that this is an artificial situation for the ants. The study of observation nests must always be treated as a supplement to work in the field, not as a substitute for it.

Estimating colony size

It is often necessary to estimate colony size. For animals as mobile as ants the mark-recapture or Lincoln Index technique would seem to be appropriate. A number of individuals are captured, alive and undamaged, and marked in some way. A known number of such marked individuals is then released back into the population. At some later time, the population is sampled at random and the number of captures and recaptures (marked individuals caught) is recorded. The total population size can then be estimated from:

$$N = \frac{Mn}{m}$$

Where:

N = estimate of population size
M = number of marked individuals released
n = total caught on second sampling occasion
m = number of marked individuals on second sampling occasion

This technique is easy to use, but time-consuming because the accuracy of the estimate increases with the number of individuals marked. It involves a number of assumptions which must be appreciated before the results are interpreted. A method of calculating 95% confidence limits for these estimates is given by Chalmers & Parker (1989).

The assumptions are:

1. The marks do not affect the animals in any way and they are not lost during the period of the investigation.
2. The marked animals become mixed within the population.
3. The population is sampled randomly.

It is the second of these that causes most problems with ants. The whole ant population does not mix freely, because some workers are restricted to the nest. Any mark-recapture exercise will only give an estimate of the numbers of foragers and this number will change with time of day, weather conditions and season. It is possible to estimate forager numbers on trails, where individual ants tend to remain on the same trail (Holt, 1955).

If the mark-recapture technique is to be used, a suitable marking method is needed. For some of the bigger species, such as the wood ants, the gaster can be marked with a small dot of quick-drying paint, such as the cellulose dope available in model shops. It does not seem to affect survival in any obvious way, but the effects of marking should be examined in specific situations, by a combination of laboratory investigations (checking such things as toxicity of marking agent) and field experiments (checking, for example, if a marked animal is more or less likely to be eaten, or to be able to catch food). Other marking methods include dusting with fluorescent powder (which shows up in ultraviolet light), powdered dyes (revealed by putting marked ants on wet white filter paper and if necessary applying a drop of acetone) or a 6% solution of phenolphthalein in acetone (which turns purple when recaptured ants are put in 1% sodium hydroxide solution), or even labelling with radioactive tracers. These have the great advantage of not being visible to any potential predator or prey, or indeed to the experimenter when performing the second capture, but their use requires special precautions.

Marking methods and the Lincoln Index technique are reviewed by Southwood (1977) and Chalmers & Parker (1989).

Territory mapping

Probably all ant species are territorial. Even if they do not show vigorous defence of an established area at all times (Hölldobler & Wilson, 1990) (an absolute territory) they will defend some parts of the foraging area (spatiotemporal territory). In either case, some assessment of territory boundaries is often needed in a study of a species or ant community. One method takes advantage of the worker's ability to recognise 'friend' or 'foe'. In wood ants, for example, workers from one colony, removed and then replaced, elicit no more than mild interest from fellow workers. In contrast, if workers from a neighbouring colony are introduced, a vigorous reaction will be observed. A 'foreign' worker will be attacked by an increasing number of 'home' workers and is eventually killed and carried off to the home nest. Similar experiments have been carried out on many other species (Hölldobler, 1979).

With care, this response can be used to map territorial boundaries quite accurately. Ants can initially be transferred from known points on the ground or within vegetation to establish which colony controls that particular spot. A further sophistication is to set out baits of 2% sugar solution in selected and known places. Aggression tests can then be used to see which, of all possible colonies, controls the place where the bait was placed. An accurate map of the study site, or some kind of grid made of string or other material, is needed for this kind of work. Techniques for mapping are described by Gilbertson and others (1985).

A simple method was use by one of the authors (GJS) for mapping trees in a wood. A tree was chosen at random on the edge of the wood and numbered 1. Compass bearings were taken with a sighting compass, and distances were measured from this tree to a few others. The data were recorded and the trees numbered as they were done. When the walking between trees got tiresome, one of the mapped trees was taken as a new base point and mapping continued from this tree. In this way, the observer can leap-frog through the wood and a large number of trees can be mapped fairly accurately and very quickly. The data can be transferred to a large sheet of graph paper with protractor and ruler. This method could be adapted to map other objects on a much smaller scale.

Feeding habits

It is fairly easy to assess the food income of a colony in trail-forming species such as *Formica* species of the wood ant group (*Formica rufa, lugubris, aquilonia, exsecta*) or *Lasius fuliginosus*. Food coming in along trails can be observed directly. A problem is, however, that if food is confiscated from a returning worker other workers become agitated in response to alarm pheromones (p. 12). Less disturbing methods have been devised. In one example (Skinner, 1980*a,b*) wood ant nests were enclosed in an ant-proof fence and the movements of foragers in and out were restricted to specific entry and exit points. The ant-proofing was achieved by burying the fence panels in a trench filled with old engine oil and by smearing its top with tree banding grease. The entrance and exits were at the tops of ramps which ended in mid-air so that incoming and outgoing flows could be split. All incoming ants fell into a box from which the only exit

Fig. 27. A Chauvin fence (see text) (from Skinner, 1980*b*).

was through holes just big enough to allow ants out. Anything they were carrying was left in the box and could be examined at leisure (fig. 27).

Another way to get information about food intake is to weigh departing and returning workers. In a study of wood ants, for example, it was shown that the mean mass of workers leaving a nest on a particular trail was significantly less than that of those returning with no food item in their mandibles, but with a swollen gaster. The difference was due to aphid honeydew in the apparently unladen returning workers. Simple weighing can provide valuable information.

Ant activity patterns

For many years ecologists and behavioural biologists have been interested in animal activity patterns and have sought to interpret them in terms of diurnal rhythms or variations related to weather, food supply and other factors. In the ants, many opportunities exist for such work. In the field activity can be investigated, simply but laboriously, by camping near a nest and making regular assessments of movements. For instance the numbers of workers leaving or entering the nest per unit time can be counted on a trail, or the number of ants visible can be counted in a known area or during a known searching period. Some of this can be automated. Automatic 'ant counters' have been devised in the past (Skinner, 1980*a*, *b*; Dibley & Lewis, 1972). The availability of modern electronic components and computer logging facilities gives scope for developing new ones. For example, a proprietary 'slot counter', used in the printing industry as a paper counter, formed the core of a laboratory ant counter. Ants from the formicarium were directed down a tube, which passed through the slot counter, and a record of each ant's passage was conveyed, by appropriate circuitry, to a desktop computer. Many days of activity could thus be logged. Specially written software was used to process the data and simple environmental control allowed some of the factors affecting activity in this species (*Lasius fuliginosus*) to be unravelled.

Effects of ants on other insects

As we have seen (p. 20), ants tend aphids and they kill and eat many other herbivorous animals, including non-tended aphids. The impact of the ants can be investigated simply by excluding ants from vegetation on which they would otherwise forage, and comparing the success of selected herbivores in this situation with their success in nearby unprotected plots. Sticky tree-banding greases, available commercially for excluding wingless moths from fruit trees (supplier p. 74), have been useful for excluding ants from trees and shrubs. Research of this kind can also be done using laboratory colonies in formicaria with large foraging arenas.

Effects on ground-living species

Ants may affect other animals living on the ground either by direct predation or by competition. Pitfall trapping (p. 64), although a technique with some problems when used for estimating numbers of ground-living animals, can be employed as a means of investigating these effects. If traps are set out around the nest, ground-living animals are caught. If their numbers are greater outside than inside a nest territory or increase with distance from the nest, predation or competition may be implicated. To find out which, if either, is operating may require further experiments or observations. For example, large counts of a particular species as incoming prey may link in with depletion of the numbers of that species near the nest. On the other hand, if a species was never brought in as prey but was much reduced in numbers near the nest, competition would be indicated. Further studies might involve supplying extra food in the ant territories to see whether the reduction of competition for food would allow the numbers of other species to increase.

Some useful addresses

Societies and special interest groups

The Bees, Wasps and Ants Recording Society (BWARS)

BWARS was inaugurated in January 1995 after functioning as a less formal scheme since 1976. The aim of the Society is to collate records of aculeates and to produce atlases giving up-to-date distribution maps and species accounts of the British aculeate fauna. BWARS will give help with identification of bees, wasps and ants and produces a biannual newsletter. Enquiries about the Society (but not specimens) should be addressed to Mr David Lloyd, BWARS Secretary, 1 Crest Road, Rochester, Kent ME1 2NG.

Royal Entomological Society, 41 Queen's Gate, London SW7 5HU
The entomologist's professional organisation, with a superb library.

International Union for the Study of Social Insects (IUSSI), c/o Dr P. J. Wright, De Montfort University, 37 Landsdowne Road, Bedford MK40 2BZ

An international society catering for the professional student of social insects. It publishes a journal called *Insectes Sociaux*.

Books, new and secondhand

E.W. Classey Ltd, P.O. Box 93, Faringdon, Oxfordshire SN7 7DR

Equipment

Watkins & Doncaster, P.O. Box 5, Cranbrook, Kent TN18 5EZ
(For all manner of equipment for the insect collector)

Hampshire Micro, The Microscope Shop, Oxford Road, Sutton Scotney, Hampshire SO21 3JG
(Suppliers of microscopes, both compound and stereo. Write for a catalogue and price list)

Interplay UK Ltd, 9A Spittal Street, Marlow, Buckinghamshire SL7 3HJ
(Suppliers of a reasonably-priced, ready-made formicarium called 'Ant World')

Vitamin capsules

Seven Seas Ltd, Marfleet, Humberside HU9 5NJ

References and further reading

Finding Books

Some of the following publications will be in local libraries, but it is likely that many will not. Your local public library may be able to arrange to borrow some items for you from the British Library. If you wish to take advantage of this service, you must be prepared to wait a number of weeks and to give full details when you make your request. The reference should be presented in full in the form given here.

References

Barrett, K.E.J. (1979). *Provisional Atlas of the Insects of the British Isles, Part 5, Hymenoptera: Formicidae. Ants*. Biological Records Centre, Institute of Terrestrial Ecology, Monks Wood Experimental Station, Huntingdon.
Baxter, F.P. & Hole, F.D. (1966). Ant (*Formica cinerea*) pedoturbation in a prairie soil. *Soil Science Society of America Proceedings* **31**, 425–428.
Beattie, A.J. (1985). *The Evolutionary Ecology of Ant–Plant Mutualisms* Cambridge: Cambridge University Press.
Berry, R.J. (1977). *Inheritance and Natural History*. London: Collins, New Naturalist.
Bhatkar, A.P. & Whitcomb, W.H. (1970). Artificial diet for rearing various species of ants. *Florida Entomologist* **53**, 229–232.
Bier, K. (1954). Uber den Einfluss der Königin auf die Arbeiterinnenfertilität im Ameisenstaat. *Insectes Sociaux* **1**, 7–19.
Billen, J.P.J. (1984). Stratification in the nest of the slave-making ant *Formica sanguinea* Latreille, 1798 (Hymenoptera, Formicidae). *Annales de la Société Royale Zoologique de Belgique* **114**, 215–225
Bolton, B. (1995). *A New General Catalogue of the Ants of the World*. Cambridge, Mass.: Harvard University Press
Bolton, B. & Collingwood, C. (1975). Hymenoptera: Formicidae. *Handbooks for the Identification of British Insects* **6** (3c). London: Royal Entomological Society of London.
Brian, M.V. (1977). *Ants*. London: Collins, New Naturalist.
Brian, M.V. (1983) *Social Insects*. London: Chapman and Hall.
Brian, M.V. & Carr, C.A.H. (1960). The influence of the queen on brood rearing in ants of the genus *Myrmica*. *Journal of Insect Physiology* **5**, 81–94.
Bristow, C.M. (1984). Differential benefits from ant attendance to two species of Homoptera on New York ironweed. *Journal of Animal Ecology* **53**, 715–726.
Buschinger, A. (1968). Mono- und Polygynie bei Arten der Gattung *Leptothorax* Mayr (Hymenoptera, Formicidae). *Insectes Sociaux* **15**, 217–226.
Carr, C.A.H. (1962). Further studies on the influence of the queen in ants of the genus *Myrmica*. *Insectes Sociaux* **9**, 197–211.
Chalmers, N. & Parker, P. (1989). *The OU Project Guide: Fieldwork and Statistics for Ecological Projects*. London: Open University/Field Studies Council.
Czerwinski, Z., Jakubczyk, H. & Petal, J. (1971). Influence of ant hills on meadow soils. *Pedobiologia* **7**, 277–285.
Darwin, C. (1859). *On the Origin of Species by Means of Natural Selection*. London: Murray.
Dibley, G.C. & Lewis, T. (1972). An ant counter and its use in the field. *Entomologia Experimentalis et Applicata* **15**, 499–508.

Disney, R.H.L. (1983). Scuttle Flies, Diptera, Phoridae (except *Megaselia*). *Handbooks for the identification of British Insects* **10**(b), 1–81. London: Royal Entomological Society of London.
Disney, R.H.L. (1994). *Scuttle Flies: the Phoridae*. London: Chapman and Hall.
Dobrzanska, J. (1959). Studies on the division of labour in ants, genus *Formica*. *Acta Biologiae Experimentalis* **19**, 57–81.
Donisthorpe, H.St.J.K. (1915). *British Ants, their Life Histories and Classification*. Plymouth: William Brendon and Son.
Donisthorpe, H. St.J.K. (1927). *The Guests of British Ants*. London: George Routledge and Son.
Douglas, J.M. & Sudd, J.H. (1980). Behavioural coordination between an aphis (*Symydobius oblongus* von Heyden; Hemiptera: Callaphidae) and the ant that attends it (*Formica lugubris* Zetterstedt; Hymenoptera: Formicidae): an ethological analysis. *Animal Behaviour* **28**, 1127–1139.
Dumpert, K. (1978). *The Social Biology of Ants* (translated by C. Johnson). London: Pitman.
Ehrhardt, S. (1931). Über Arbeitslung bei *Myrmica*- and *Messor*-arten. *Zeitschrift für Morphologie und Ökologie der Tiere* **20**, 755–812.
Elgert, B. & Rosengren, R. (1977). The guest ant *Formicoxenus nitidulus* follows the scent trail of its wood ant host (Hymenoptera, Formicidae). *Memoranda Societatis pro Fauna et Flora Fennica* **53**, 35–38.
Elton, C. (1966). *The Pattern of Animal Communities*. London: Methuen.
El-Ziady, S. & Kennedy, J.S. (1956). Beneficial effects of the common garden ant, *Lasius niger* L., on the black bean aphid, *Aphis fabae* Scopoli. *Proceedings of the Royal Entomological Society of London* A, **31**, 61–65.
Ettershank, G. (1967). A completely defined synthetic diet for ants (Hym., Formicidae). *Entomologist's Monthly Magazine* **103**, 66–67.
Feynmann, R.P. (Hutchings, E. (ed.)) (1985). *Surely You're Joking, Mr. Feynmann*. New York: Morton.
Franks, N.R., Healy, K.J. & Byrom, L. (1990). Studies on the relationship between the ant ectoparasite *Antennophorus grandis* (Acarina: Antennophoridae) and its host *Lasius flavus* (Hymenoptera: Formicidae). *Journal of Zoology* **225**, 59–70.
Gauld, I. & Bolton, B. (eds.) (1988). *The Hymenoptera*. London: British Museum (Natural History).
Gilbertson, D.D., Kent, M. & Pyatt, F.B. (1985). *Practical Ecology for Geography and Biology*. London: Unwin Hyman.
Gösswald, K. (1985). *Organisation und Leben der Ameisen*. Stuttgart: Wissenschaftliche Verlagsgesellschaft MBH.
Gösswald, K. & Bier, K. (1953). Untersuchungen zur Kastendetermination in der Gattung *Formica*. 2. Die Aufzucht von Geschlechtstieren bei *Formica rufa pratensis* (Retz). *Zoologischer Anzeiger* **151**, 126–134.
Gould, J.L. & Gould, C.G. (1988). *The Honey Bee*. New York: Scientific American Library.
Haldane, J.B.S. (1955). Population genetics. *New Biology* **18**, 34–51.
Hamilton, W.D. (1964). The genetical evolution of social behaviour. *Journal of Theoretical Biology* **7**, 1–52.
Hamilton, W.D. (1972). Altruism and related phenomena, mainly in social insects. *Annual Review of Ecology and Systematics* **3**, 193–232.
Hangartner, W. (1967). Spezifität und Inaktivierung des Sphurpheromons von *Lasius fuliginosus* Latr. und Orientierung der Arbeiterinnen im Duftfeld. *Zeitschrift für Vergleichende Physiologie* **57**, 103–136.

Heyde, K. (1924). Die Entwicklung der psychischen Fähigkeiten bei Ameisen und ihr Verhalten bei abgeänderten biologischen Bedigungen. *Biologisches Zentralblatt* **44**, 623–654.

Hickman, J.C. (1974). Pollination by ants: a low energy system. *Science* **184**, 1290–1292.

Hölldobler, B. (1979). Territories of the African Weaver ant (*Oecophylla longinoda*): a field study. *Zeitschrift für Tierpsychologie* **51**, 201–213.

Hölldobler, B. & Wilson, E.O. (1990). *The Ants.* Berlin: Springer-Verlag.

Hölldobler, K. (1953). Beobachtungen über die koloniegründung von *Lasius umbratus* Nyl. *Zeitschrift für angewandte Entomologie* **34**, 598–606.

Holt, S.J. (1955). On the foraging activity of the wood ant. *Journal of Animal Ecology* **24**, 1–34.

Horstmann, K. & Schmidt, H. (1986). Temperature regulation in nests of the wood ant *Formica polyctena* (Hymenoptera: Formicidae). *Entomologia Generalis* **11**, 229–236.

Huber, P. (1810). *Recherches sur les Moeurs des Fourmis Indigenes.* Paris: J.J. Paschoud.

Janzen, D.H. (1967). Interaction of the bull's-horn acacia (*Acacia cornigera* L.) with an ant inhabitant (*Pseudomyrmex ferruginea* F. Smith) in eastern Mexico. *University of Kansas Science Bulletin* **47**, 315–558.

King, T.J. (1977). The plant ecology of ant-hills in calcareous grasslands. I. Patterns of species in relation to ant-hills in southern England. *Journal of Ecology* **65**, 235–256.

Kleinjan, J.E. & Mittler, T.E. (1975). A chemical influence of ants on wing development in aphids. *Entomologia Experimentalis et Applicata* **18**, 384–388

Kloft, W.J. (1959). Versuch einer Analyse der trophobiotischen Beziehungen von Ameisen und Aphiden. *Biologisches Zentralblatt* **78**, 863–870.

Kondoh, M. (1968). Bioeconomic studies on the colony of an ant species, *Formica japonica* Motschulsky. 1. Nest structure and seasonal change of the colony members. *Japanese Journal of Ecology* **18**, 124–133.

Koptur, S. & Lawton, J.H. (1988). Interactions among vetches bearing extrafloral nectaries, their biotic protection agents, and herbivores. *Ecology* **69**, 278–283.

Kunkel, H. (1973). Die Kotabgade der Aphiden (Aphidina, Hemiptera) unter Einfluss von Ameisen. *Bonner Zoologische Beitrage* **24**, 105–121.

MacGregor, E.C. (1948). Odour as a basis for orientated movement in ants. *Behaviour* **1**, 267–296.

Majerus, M. & Kearns, P. (1989). *Ladybirds.* Naturalists' Handbooks 10. Slough: The Richmond Publishing Co. Ltd.

Moffett, M.W. (1987). Ants that go with the flow: a new method of orientation by mass communication. *Natturwissenschaften* **74**, 551–553.

Nielsen, M.G. (1972). An attempt to estimate energy flow through a population of workers of *Lasius alienus* (Först) (Hymenoptera: Formicidae). *Natura Jutlandica* **16**, 99–107.

Nonacs, P. (1986). Ant reproductive strategies and sex allocation theory. *Quarterly Review of Biology* **61**, 1–21.

Otto, D. (1958). Über die Arbeitsteilung im Staate von *Formica rufa rufo-pratensis minor* Gössw. und ihre verhaltenphysiologischen Grundlagen: ein Beitrag zur Biologie der Roten Waldameise. *Wissenschaftliche Abhandlungen der Deutschen Akademie der Landwirtschaftwissenschaften zu Berlin* **30**, 1–169.

Peakall, R., Handel, S.N. & Beattie, A.J. (1991). The evidence for, and importance of, ant pollination. In Huxley, C.R. & Cutler, D. F (eds.) *Ant-Plant Interactions.* Oxford: Oxford University Press.

Rashbrook, V.K., Compton, S.G. & Lawton, J.H. (1991). Bracken and ants: why is there no mutualism? In Huxley, C.R. & Cutler, D.F. (eds.) *Ant-Plant Interactions.* Oxford: Oxford University Press.

Rosengren, R. (1987). Polyethic structure of the foraging/guarding system of red wood ants (*Formica s.* str.). In Eder, J. and Rembold, H. (eds.) *Chemistry and Biology of Social Insects.* Proceedings of the Tenth International Congress of the International Union for the Study of Social Insects, Munich, 1968. pp. 118–119. Munich: J. Peperny.

Rotheray, G. (1989). *Aphid Predators.* Naturalists' Handbooks 11. Slough: The Richmond Publishing Co. Ltd.

Rotheray, G. (1994). *Insect Life on Plants.* London: Chapman and Hall.

Salem, M. & Hole, F.D. (1968). Ant (*Formica cinerea*) pedoturbation in a forest soil. *Soil Science Society of America Proceedings* **32**, 563–567.

Scherba, G. (1959). Moisture regulation in mound nests of the ant, *Formica ulkei* Emery. *American Midland Naturalist* **61**, 499–508.

Seeley, T.D. (1985). *Honeybee Ecology.* Princeton, N.J.: Princeton University Press.

Seifert, B. (1992). A taxonomic revision of the Palaearctic members of the ant subgenus *Lasius s.* str. (Hymenoptera: Formicidae). *Abhandlungen und Berichte des Naturkundemuseums Görlitz* **66** (5), 1–67.

Shaw, M.R. & Huddleston, T. (1991). Classification and biology of braconid wasps (Hymenoptera: Braconidae). *Handbooks for the Identification of British Insects,* **7** (11). London: Royal Entomological Society.

Skaife, S.H. (1961). *The Study of Ants.* London: Longmans Green.

Skinner, G.J. (1980a). Territory, trail structure and activity patterns in the wood-ant *Formica rufa* (Hymenoptera: Formicidae) in limestone woodland in north-west England. *Journal of Animal Ecology* **49**, 381–394.

Skinner, G.J. (1980b). The feeding habits of the wood ant *Formica rufa* (Hymenoptera, Formicidae) in limestone woodland in north-west England. *Journal of Animal Ecology* **49**, 417–433.

Skinner, G.J. (1987). *Ants of the British Isles.* Princes Risborough: Shire Publications Ltd.

Skinner, G.J. & Whittaker, J.B. (1981). An experimental investigation of inter-relationships between the wood-ant (*Formica rufa*) and some tree-canopy herbivores. *Journal of Animal Ecology* **50**, 313–326.

Southwood, T.R.E. (1977). *Ecological Methods.* London: Chapman and Hall.

Sudd, J.H. (1967). *An Introduction to the Study of the Behaviour of Ants.* London: Edward Arnold.

Sudd, J. & Franks, N. (1987). *The Behavioural Ecology of Ants.* London: Chapman and Hall.

Thomas, J. & Lewington, R. (1991). *The Butterflies of Britain and Ireland.* London: Dorling Kindersley.

Vitzthum, H. (1929). Milben, Acari. In *Die Tierwelt Mitteleuropas* **3** (7), 1–112. Leipzig: Quelle und Meyer.

Vitzthum, H. (1940–43). Acarina. In Bronn's *Klassen und Ordnungen des Tierreichs.* Leipzig, pp. 1–1011.

Way, M.J. (1954). Studies of the association of the ant *Oecophylla longinoda* (Latr.) (Formicidae) with the scale insect *Saissetia zanzibarensis* Williams (Coccidae). *Bulletin of Entomological Research* **45**, 113–134.

Weir, J.S. (1958a). Polyethism in workers of the ant *Myrmica.* I. *Insectes Sociaux* **5**, 97–128.

Weir, J.S. (1958b). Polyethism in workers of the ant *Myrmica* II. *Insectes Sociaux* **5**, 315–339.

Wheeler, W.M. (1910). *Ants: their Structure, Development and Behaviour.* New York: Columbia University Press.

White, W.F. (1895). *Ants and their Ways*. London: The Religious Tract Society.

Wiken, E.B., Broersma, K., Lavkulich, L.M. & Farstad, l. (1976). Biosynthetic alteration in a British Columbia soil by ants (*Formica fusca* Linne.). *Soil Science Society of America Journal* **40**, 422–426.

Wilson, E.O. (1971). *The Insect Societies*. Cambridge, Mass.: Harvard University Press.

Wilson, E.O. (1975). *Sociobiology, the New Synthesis*. Cambridge, Mass.: Belknap/Harvard.

Yeo, P.F. & Corbet, S.A. (1995). *Solitary wasps*. Naturalists' Handbooks 3. Slough: The Richmond Publishing Co. Ltd.

Zahn, M. (1958). Temperatursinn, Wärmehaushalt und Bauweise der roten Waldameisen (*Formica rufa* L.) *Zoologische Beitrage*, n.s. **3**, 127–194.

Index

abdomen 38
Acropyga species 21
activity pattern 72
   daily variation 3
   seasonal variation 3
adonis blue 33
Aenigmatias 31
aggression 12
alarm signal 20
alitrunk 38
Anergates atratulus 15, 50, 55, 61, pl. 8.3, pl. 10.6
   A. species 3, 34, 36
ant counter 72
ant guest 29–31
antenna 39
Antennophorus grandis 31
antibiotic 28
aphid 20–23
   sycamore 21
Aphis species 22
appeasement behaviour 30
aspect 9
aspirator 64
Atemeles emarginatus 31, pl. 4.5
   A. paradoxus 31

bait 70
beating 65
behaviour 22
blue butterfly 29, 33
bracken 20, 27
brood 8
butterfly 29, 33

Callophrys rubi 33
Calluna vulgaris 15
caruncle 24
caste 10, 39
   determination 12, 13, 18
   sub-caste 11
   temporal 11
chalkhill blue 33
chemical communication 23
claustral 16
Claviger longicornis 31
   C. testaceus 30, 31, pl. 4.4
cleptobiosis 34
clypeus 39
collection 64
colony 10
   adolescent 18, 19
   founding 17
   growth 17
   juvenile 18, 19
   mature 18, 19
   odour 10
   size 69
common blue 33
community 14, 15
competition, intra-specific 15
compound eye 39
cornicles 21
coxa 38
culture 67

data logging 4
desiccation 9
diapause 13
diet 12, 20, 68
Dinarda species 30
diploidy 12
dispersal 24
   of plants 28
distribution 3
division of labour 10
Dolichoderinae 2, 40, 47, 53
domatia 27
Doronomyrmex species 3
Drepanosiphum platanoidis 21
dulosis 34

egg 11–13, 15, 16, 18
   trophic 16
   winter 14
energy input, daily variation 3
   seasonal variation 3
excavation 7
extrafloral nectaries 27

feeding 19, 71
female calling syndrome 16
femur 38
finding ants 64
food 19–24
   intake, daily variation 3
   intake, seasonal variation 3
   seed 24
foraging strategy 25
Forda formicaria 21
formic acid 12, 37, 39
Formica aquilonia 42, 48, 54, 58, 64, pl. 6.4
   F. candida (=*transkaucasica*) 15, 41, 48, 54, 65
   F. cunicularia 15, 42, 48, 54, 65, pl. 1.4
   F. exsecta 17, 41, 47, 53, 58, 64, pl. 1.5, pl. 10.9

Index

F. fusca 7, 17, 31, 32, 41, 48, 54, 58, 64, pl. 6.5
F. japonica 8
F. lemani 11, 41, 48, 54, 58, 64
F. lugubris 17, 42, 48, 54, 58, 64, pl. 1.1
F. nigricans 12
F. polyctena 9, 12
F. pratensis 42, 48, 54, 64
F. rufa 11, 16, 17, 27, 35, 42, 48, 54, 58, 64, pl. 1.2, pl.1.3, pl. 1.6
F. rufibarbis 42, 48, 54, 65
F. sanguinea 11, 31, 32, 34, 35, 41, 47, 53, 58, 64, pl. 2.1, pl. 2.2
F. species 3, 8, 41, 47, 53
F. ulkei 9
formicarium 3, 7, 67
Formicidae 1
Formicinae 2, 40, 47, 53
Formicoxenus nitidulus 3, 25, 34, 35, 45, 51, 55, 59, 61, pl. 5.3
frons 39
funiculus 39

gaster 38, 39
genital capsule 39
green hairstreak 33
growth curve 17
guest ant 35
guests
    intranidal 29, 30
    extranidal 33

haploidy 12, 16
head 38, 39
    width 11
honeydew 12, 20–23
humidity 9, 10
Hypoponera punctatissima 2, 40, pl. 2.5

inclusive fitness 6
inquilinism 34, 36
insulation 9
Iridomyrmex humilis 2

juvenile hormone (JH) 22

killing ants 65
kin selection 5

Lachnus species 23
large blue 33, pl. 4.6
larva 9, 13, 14, 18
Lasius alienus 7, 31, 43, 49, 55, 65
    L. brunneus 27, 43, 50, 55, 65

L. flavus 3, 15, 31, 32, 37, 43, 50, 55, 59, 65, pl. 9.2, pl. 9.3
L. fuliginosus 11, 17, 25, 26, 31, 32, 42, 49, 55, 58, 65, pl. 2.3, pl. 9.3, pl. 10.10
L. meridionalis 41, 42, 43, 49, 55
L. mixtus 31, 32, 43, 49, 55, 65
L. niger 7, 15, 25, 31, 32, 39, 43, 49, 55, 59, 65, pl.4.1, pl. 4.2, pl. 9.1
L. reginae 17
L. sabularum 42, 49, 54
L. species 32, 38, 41, 47, 53
L. umbratus 17, 32, 42, 43, 49, 55, 65, pl. 2.4
Leptothorax acervorum 11, 45, 51, 56, 61, pl. 5.2, pl, 10.7
L. interruptus 27, 45, 51, 57, pl. 3.1
L. nylanderi 11, 27, 45, 51, 57
L. species 27, 45, 51, 57, 59, 61
L. tuberum 38, 45, 51, 57, pl. 5.1
L. unifasciatus 12
lestobiosis 34
Lincoln Index 69
Linepithema humile 40
ling 15
Lomechusa strumosa 31
Lycaenidae 33
Lysandra bellargus 33
    L. coridon 33

Maculinea arion 33, pl. 4.6
male 10, 11, 53
    aggregation 16
mandible 39
marking 11
    technique 69, 70
maxillary gland 12
mesopleuron 38
mesoscutellum 38
metatarsus 38
Methocha fimbricornis 5
microclimate 8
microscope 39
mites 32
monogyny 17
moult 15
mounting 65
Myrmecina graminicola 44, 50, 57, 61, pl. 7.3, pl. 10.3
myrmecophily 29
Myrmedonia humeralis 30
Myrmica bessarabica (=*specioides*) 46, 52, 57
    M. hirsuta 3, 34, 36, 51, 56, pl. 3.5
    M. karavajevi 3, 34, 36, 51, 56

Index

M. lobicornis	32, 46, 52, 57, 60
M. rubra	11, 15, 32, 45, 52, 56, 60, pl. 8.1, pl. 10.5
M. ruginodis	15, 18, 25, 26, 32, 45, 52, 56, 60, pl. 3.2, pl. 3.4, pl. 3.5
M. sabuleti	3, 15, 46, 52, 57, 60
M. scabrinodis	3, 15, 31, 32, 38, 46, 52, 57, 60, pl. 8.2
M. schenki	46, 52, 57, 60
M. species	14, 25, 44, 51, 56, 59, 60, pl. 4.3
M. sulcinodis	45, 52, 56, 60, pl. 3.3
Myrmicinae	2, 40, 47, 53
navigation	25
nectar	18
nest	3, 7–10, 65
mound	7, 8
thatched	6
nuptial flight	16
odour	10
ocellus	39
Oecophylla longinoda	22
palps	39
parabiosis	34
parasite	3, 22
ectoparasite	29, 31
endoparasite	29, 31
temporary social	17, 34
permanent	34
social	34, 51
parasitoid	29, 32
parental manipulation	5
pedicel	38
Periphyllus testudinaceus	21
pest	27
petiole	38
pheasant	34
Pheidole tepicana	11
pheromone	13
sex	16
Picus viridis	10, 33
pitfall trap	64
plants	26–29
ant dispersal of	28
ant pollination of	28
ant pruning of	28
as food	27
as shelter	27
Plebejus argus	33
plesiobiosis	34
poison gland	39
pollination	28
polyethism, age-specific	11, 12
Polyommatus icarus	33
Ponera coarctata	2, 40, 47, 53, 60, pl. 7.1, pl. 7.2, pl. 10.1
Ponerinae	2, 40, 47, 53
pooter	64
postpetiole	38
Potentilla erecta	24, 28
predator	22, 23
preserving ants	64
prey	4, 20, 23, 24
pronotum	38
propodeum	38
Pteridium aquilinum	20, 27
pupa	9
purple hairstreak	33
queen	10–12, 16, 17, 47
queen substance	13
Quercusia quercus	33
records	3
recruitment	25
rove beetle	30
Saissetia zanzibarensis	22
sampling device	18
scape	39
scent trail	25
scutellum	38
seed	24
sexing ants	39
silver-studded blue	33
slavery	34, 35
social behaviour	1, 5, 6
soil	36
aeration of	17
chemical effect of ants	37
enrichment of	27
pH	37, 38
physical effect of ants	37
Solenopsis fugax	3, 34, 35, 43, 50, 56, 61, pl. 6.2
S. invicta	17
Sphecomyrma freyi	4, 5
spider	33
spiracle	38
Staphylinidae	30
Stenamma debile	38, 44, 50, 57
S. species	44, 50, 57
S. westwoodii	44, 50, 57, 61
sticky grease	22
sting	39
storage of collection	66
Strongylognathus species	3, 36

S. testaceus	15, 34–36, 43, 50, 56, 61, pl. 6.1, pl. 10.4
swarm	16
sweep net	64
symphile	29, 30
Symydobius oblongus	22
synechthran	29, 30
synoekete	29, 30
tandem running	25,
Tapinoma ambiguum	40, 47, 53, pl. 6.3
T. erraticum	2, 15, 40, 47, 53, 59, 61, pl. 10.8
T. melanocephalum	2, 40
T. species	40, 47, 53
temperature	9
territory	10,
mapping	70
Tetramorium caespitum	15, 25, 32, 36, 44, 50, 56, 59, 61, pl. 2.6, pl. 5.4, pl. 10.2
Theridiidae	33
thief ant	35
thorax	38
tibia	38
tiphiid wasp	5
trail	19, 25
tramp species	2, 40
trochanter	38
trophobiont	29
trophobiotic organ	22
Ulex species	2
Viola species	28
waist	1, 38
wing muscle	16
wood ant	7, 11, 19, 20, 41, 48
woodpecker	10, 33
worker	5, 10, 11, 17
major	11
media	11
minor	11
xenobiosis	34